# REASONING IN BOOLEAN NETWORKS

## Logic Synthesis and Verification Using Testing Techniques

# FRONTIERS IN ELECTRONIC TESTING

*Consulting Editor*
**Vishwani D. Agrawal**

***Books in the series:***

**Introduction to $I_{DDQ}$ Testing**
S. Chakravarty, P.J. Thadikaran
ISBN: 0-7923-9945-5
**Multi-Chip Module Test Strategies**
Y. Zorian
ISBN: 0-7923-9920-X
**Testing and Testable Design of High-Density Random-Access Memories**
P. Mazumder, K. Chakraborty
ISBN: 0-7923-9782-7
**From Contamination to Defects, Faults and Yield Loss**
J.B. Khare, W. Maly
ISBN: 0-7923-9714-2
**Efficient Branch and Bound Search with Applications to Computer-Aided Design**
X.Chen, M.L. Bushnell
ISBN: 0-7923-9673-1
**Testability Concepts for Digital ICs: The Macro Test Approach**
F.P.M. Beenker, R.G. Bennetts, A.P. Thijssen
ISBN: 0-7923-9658-8
**Economics of Electronic Design, Manufacture and Test**
M. Abadir, A.P. Ambler
ISBN: 0-7923-9471-2
**$I_{DDQ}$ Testing of VLSI Circuits**
R. Gulati, C. Hawkins
ISBN: 0-7923-9315-5

# REASONING IN BOOLEAN NETWORKS

## Logic Synthesis and Verification using Testing Techniques

by

**Wolfgang Kunz**
University of Potsdam, Germany

and

**Dominik Stoffel**
University of Potsdam, Germany

**Kluwer Academic Publishers**
Boston/Dordrecht/London

A C.I.P. Catalogue record for this book is available from the Library of Congress

ISBN 0-7923-9921-8

---

Published by Kluwer Academic Publishers,
P.O. Box 17, 3300 AA Dordrecht, The Netherlands.

Sold and distributed in the U.S.A. and Canada
by Kluwer Academic Publishers,
101 Philip Drive, Norwell, MA 02061, U.S.A.

In all other countries, sold and distributed
by Kluwer Academic Publishers Group,
P.O. Box 322, 3300 AH Dordrecht, The Netherlands.

*Printed on acid-free paper*



Printed in the Netherlands

"Logic, old or new, traces implications."

(W.V.O. Quine, 1981, in *Theories and Things*)

# CONTENTS

**FOREWORD** XI

**PREFACE** XIII

**1 PRELIMINARIES** 1

1.1 Boolean Algebra 1

1.2 Graphs 4

1.3 Boolean Functions and their Representations 5
1.3.1 Implementing Boolean Functions as Switching Circuit 6
1.3.2 Disjunctive Forms 7
1.3.3 Boolean Networks 10
1.3.4 Binary Decision Diagrams 11

**2 COMBINATIONAL ATPG** 17

2.1 Problem Formulation 17

2.2 The Decision Tree 21

2.3 Logic Value Alphabets and Implications 23

2.4 General Procedure 28

2.5 Topological Analysis for Unique Sensitization 35

2.6 The Problem of ATPG: An Example 38

2.7 Static and Dynamic Learning 42

**3 RECURSIVE LEARNING** **49**

3.1 Example 50

3.2 Determining all necessary assignments 55
3.2.1 The Complete Implication Procedure 55
3.2.2 The Complete Unique Sensitization Procedure 63

3.3 Test Generation with Recursive Learning 67
3.3.1 An Algorithm to Choose the Maximum Recursion Depth $r_{max}$ 67
3.3.2 Experimental Results 69

**4 AND/OR REASONING GRAPHS** **75**

4.1 OR Search versus AND/OR Search 77

4.2 AND/OR Reasoning Trees 82

4.3 Implicants in Multi-Level Circuits 90
4.3.1 Prime Implicants 91
4.3.2 Permissible Prime Implicants 94

4.4 Search State Hashing and Isomorphic Subtrees 97

**5 LOGIC OPTIMIZATION** **101**

5.1 Problem Formulation 101

5.2 Functional Decomposition 103

5.3 Boolean and Algebraic Methods 108
5.3.1 Division 108
5.3.2 Kernel Extraction 113
5.3.3 Optimization Strategies 116
5.3.4 Internal Don't Cares 119

5.4 Global Flow 120

5.5 Redundancy Elimination 125

5.6 Implication-Based Methods 129

5.6.1 Transforming Combinational Circuits by ATPG and Implications 131
5.6.2 Heuristics to Select Implicants 140
5.6.3 Optimization Procedure 152
5.6.4 Experimental Results 154

**6 LOGIC VERIFICATION 163**

6.1 Motivation 163

6.2 A Hybrid Approach to Logic Verification 166

6.3 Exploiting Structural Properties in Verification 169
6.3.1 Exploiting Structural Similarity by Storing of Implications 171
6.3.2 Exploiting Structural Similarity by Miter Optimization 175

6.4 Functional Comparison and Solving False Negatives 181
6.4.1 Pseudo-Input Justification 183
6.4.2 Incorporating Don't Cares 185
6.4.3 Discussion of Possible Extensions to Functional Phase 189

6.5 Experimental Results 190

**7 CONCLUSIONS AND FUTURE WORK 197**

**REFERENCES 201**

**APPENDIX 213**

**INDEX 227**

# FOREWORD

Since the development of electronic logic circuits over 50 years ago, researchers have formulated numerous techniques for analyzing their behavior. Typically, a technique is targeted to a particular application, such as circuit optimization, test generation, or logic verification, but often a technique developed for one problem can be applied to others. Much of the research effort has focussed on combinational circuits consisting of acyclic networks of logic gates.

One might expect after all these years of research on combinational gate networks, that the field would have "dried up". Either techniques would have been developed that effectively solve all of the problems or progress would have come to a standstill short of this point. Interestingly, neither of conditions has occurred. The sizes of the networks to analyze, often containing thousands of gates, and the fact that the problems are inherently intractable (assuming that P does not equal NP) ensure that challenges will always remain. On the other hand, genuine progress is being made. For example, tests can be generated for circuits today that would far exceed the capability of earlier tools, even discounting the advantage of today's more powerful computers.

Every few years, a new idea comes along that clearly advances the state of the art in circuit analysis. Such is the case for recursive learning, originally presented by Kunz and Pradhan in 1992. Building on the combinatorial search approach used by most automatic test pattern generators, recursive learning stores information about the relationships between signals derived during the search. This stored information can then vastly improve the performance of subsequent search operations. Recursive learning has proved a powerful technique for a variety of application domains, including test generation, circuit equivalence testing, and logic optimization.

This book provides a comprehensive treatment of combinational network analysis, with a particular focus on recursive learning and its applications. The extended format of a book allows a thorough and self contained treatment of the subject. It would be difficult to extract this information from the numerous conference and journal papers that have been published in this area. The authors have done a good job of describing not just their own work, but also that of their predecessors. The net result is a very readable and informative book.

Pittsburgh, Pennsylvania, USA<br>May 14, 1997

Randal E. Bryant<br>Carnegie Mellon University

# PREFACE

Algorithms for logic synthesis and VLSI testing have been under investigation for several decades. For many problems in this area, it is a central issue to have appropriate ways of handling Boolean functions. A lot of research has recently been focused on this point. Important research developments have been triggered especially by the introduction of *binary decision diagrams* (BDDs) paving the way towards efficient procedures for solving many design automation problems. On the other hand, needless to say, modern representations of Boolean functions cannot overrule the results of complexity theory and practical experiences show that in spite of exciting progress in many areas there is still a number of applications where currently available methods do not provide satisfactory performance on today's industrial designs.

In this book we develop new methods for analyzing the logic function of a given circuit. We describe an alternative approach to deal with Boolean functions and their implementations as multi-level circuits that promises to be useful for supplementing existing Boolean representations in cases where they fail. In our view, binary decision diagrams and related Boolean representations have mainly two shortcomings. The first problem is that they are *exhaustive*, i.e., they are useful only if the function is fully represented. This means that there will always exist classes of circuits that cannot be handled exclusively by these representations. The second disadvantage is that they are hard to use for *systematic reasoning*. Especially this point is of great interest to us. The goal of any reasoning is to derive the logic consequences of a given assumption. Specifically, for some Boolean statement $A$ we would like to derive some statement $B$ that is true if $A$ is true, i.e., $A \Rightarrow B$. Previous representations of Boolean functions are not well suited for this kind of task. For example, given statement $A$, a BDD-based approach cannot *derive* or *imply* statement $B$, it can only *check* if $A \Rightarrow B$ is true when both $A$ and $B$ are given.

In the past, methods for Boolean reasoning have played a more important role in the research of design automation than they do today. Since the work of Quine and McCluskey the notion of *implicants* is central in every undergraduate-level course on logic design and is an important element in the theory of two-level minimization. However, for *multi-level* circuits this notion has been used only rarely to formulate effective procedures. The reason is that the classical concepts are tailored for the special structure of two-level circuits and do not appropriately take into account the structural properties of general multi-level combinational networks. In this book, we have extended the notion of implicants to multi-level circuits (see Section 4.3) and examined the role that this basic concept can play in algorithms for solving various design automation problems. We describe specific *reasoning techniques* (Chapter 3 and 4) for multi-level circuits to identify implications and implicants. This leads to the representation of Boolean functions by *AND/OR reasoning graphs* (Chapter 4). These graphs can fully represent a Boolean function but they are also useful when it is intractable to build them completely. Rather than representing and manipulating the complete functions our techniques look at important elements of these functions, namely implicants or sets of implicants. Based on the notion of implicants generalized for multi-level circuits this book develops new procedures for test generation (Chapter 3), synthesis (Chapter 5) and verification (Chapter 6).

A second objective of this book is to demonstrate the usefulness of testing techniques in logic synthesis. Research advances in the fields of VLSI testing and logic synthesis usually have been achieved independently of the developments in the other area. Only in recent years efforts have been made to investigate the interrelationships between the two fields. The driving force to bridge the gap between the two research domains came from the need to develop *synthesis for testability* methodologies. A lot of research has been conducted to relate the effect of specific synthesis procedures on the testability properties of a design. Some circuit properties required for testing can be related to common concepts of logic synthesis. For example, it is well-known that prime and irredundant circuits are 100% testable for single stuck-at faults. However, up to now, with only few exceptions, *algorithmic* issues have been considered separately for the two domains. While it is common to argue that modern synthesis techniques lead to designs that make test generation a fairly easy task (so that complex test generation procedures become superfluous), this book is dedicated to the other side of the coin. It is attempted to show that the algorithmic concepts of modern ATPG (Automatic Test Pattern Generation) have applications far beyond test generation and can represent core techniques in logic synthesis and formal verification.

This book is mainly a report of our recent research developments and does not intend to give full and detailed coverage of all relevant techniques that have evolved in these areas. Nevertheless, we hope that the book can also serve as an introduction into the algorithmic principles of test generation and logic synthesis and that it will contribute to deepen the understanding of the intimate relationship between these areas.

Most of the described research has been conducted at the Max-Planck Fault-Tolerant Computing Group at the University of Potsdam and at the Institut für Theoretische Elektrotechnik at the University of Hannover, Germany. First of all, the authors wish to thank the heads of these two research institutions, Michael Gössel and Joachim Mucha. Their personal encouragement, scientific suggestions as well as the continual effort to provide excellent working environments have been invaluable for this work. Furthermore, we especially thank Dhiraj Pradhan, Texas A&M University, and Prem Menon, University of Massachusetts, who participated in several projects and co-authored papers with us on this subject.

For his guidance and valuable suggestions during the peparation of this manuscript we are very grateful to the consulting editor, Vishwani Agrawal, Bell Labs, Lucent Technologies.

Over the years, several researchers have supported us in various ways, co-authored papers with us and contributed to this research. We thank Mitrajit Chatterjee, Martin Cobernuss, Stefan Gerber, Torsten Grüning, Sandeep Gupta, Wuudiann Ke, Udo Mahlstedt and Subodh Reddy. Thanks to their active participation, this research has been both, fruitful and very enjoyable.

Potsdam, December 1996

Wolfgang Kunz
Dominik Stoffel

# Chapter 1

# PRELIMINARIES

This chapter briefly reviews the basic concepts of switching theory. This summary is not complete; only such topics are considered which are of relevance for the understanding of later chapters. For a more detailed and rigorous introduction into the theory of switching functions the reader may refer to a standard text book, e.g., [Koha78], [McCl86] and [Fabr92]. A second objective of this chapter is to familiarize the reader with the symbolic notations used in later chapters of this book.

## 1.1 Boolean Algebra

In the following, Boolean algebra is introduced as a complementary and distributive lattice. Standard notations are used to describe sets and operations on sets.

An *ordered pair* $(a, b)$ is a pair of two elements with an order associated with them. The *Cartesian product* of two sets $A$ and $B$, denoted $A \times B$, is the set of all ordered pairs $(a, b)$, such that $a \in A$ and $b \in B$. A subset $R$ of $A \times B$ is called a *binary relation* between $A$ and $B$ and we denote $a\ R\ b$ to express that element $a$ is related to $b$ by $R$. Since only *binary* relations are considered here, for reasons of brevity we simply speak of *relations*.

The following properties of relations are important: A relation $R$ between two sets $A$ and $B$ is called

i) *reflexive*, if $(a, a) \in R$
ii) *symmetric*, if $(a, b) \in R \Rightarrow (b, a) \in R$
iii) *antisymmetric*, if $(a, b) \in R$ and $(b, a) \in R \Rightarrow a = b$
iv) *transitive*, if $(a, b) \in R$ and $(b, c) \in \mathrm{R} \Rightarrow (a, c) \in R$.

An *equivalence relation* is a relation that is reflexive, symmetric and transitive. If a relation is reflexive, antisymmetric and transitive it is called a *partial order.* A set $S$ in combination with a partial order is called a *partially ordered set.*

Let $R$ be a partial order and $S$ a partially ordered set. The reader may think of relation $R$ as the "is less than or equal to" relation to better understand the following notions. If and only if (iff) $a\ R\ b$ for every element $b \in S$, then $a$ is called the *least element* of $S$. Similarly, $a$ is said to be the *greatest element* of $S$, iff $b\ R\ a$ for all $b \in S$. Further let $P$ be a subset of $S$. An element $s \in S$ is called an *upper bound* of $P$ iff, for every $p \in P$, $p\ R\ s$. It is called a *lower bound* of $P$ if and only if, for every $p \in P$, $s\ R\ p$. An upper bound $s$ of $P$ is defined to be the *least upper bound* iff $s\ R\ s'$ for all upper bounds $s'$ of $P$. Similarly, a lower bound $s$ of $P$ is called the *greatest lower bound* iff $s'\ R\ s$ for all lower bounds $s'$ of $P$.

The notions of upper and lower bounds become intuitively clear if a partially ordered set is represented by a *Hasse diagram* as shown in standard text books. In the Hasse diagram, a partially ordered set for which a unique least upper bound and a unique greatest lower bound exists for every pair of elements looks like a grid or *lattice*.

**Definition 1.1:** A *lattice* is a partially ordered set where every pair of elements has a unique greatest lower bound and a unique lowest upper bound.

It immediately follows from this definition that a lattice has both, a least and a greatest element. We denote the least element by 0 and the greatest element by 1. Determining the least upper bound and the greatest lower bound can be viewed as two operations on the elements of the lattice. The operations consist in assigning the unique lowest upper bound or the unique greatest upper bound to each ordered pair of elements. These operations, in the following, are called *sum* or *product*, respectively, and are denoted:

$$a + b = \text{lowest upper bound } (a, b)$$
$$a \cdot b = \text{greatest lower bound } (a, b)$$

Based on the these operations a Boolean algebra can be defined as follows.

**Definition 1.2:** A lattice is called a *Boolean algebra* iff it is *complemented* and *distributive*, i.e., if the lattice fulfills the following two conditions:

*distributivity*: $a \cdot (b + c) = a \cdot b + a \cdot c$ and $a + (b \cdot c) = (a + b) \cdot (a + c)$
*complement*: for each element $a$ in the lattice there exists a unique element $\bar{a}$ such that $a \cdot \bar{a} = 0$ and $a + \bar{a} = 1$

Since the complement is unique, determining the complement can be considered a third operation on the elements of the lattice.

We have introduced a Boolean algebra as a special lattice. Another common way to define a Boolean algebra is to postulate certain properties for a set $S$ and two binary operations $(\cdot)$ and $(+)$. These properties are known as *Huntington's postulates*. They include the existence of a complement and distributivity as given in Definition 1.2. Further postulates are *idempotency*, *commutativity*, *absorption*, *associativity* and the existence of *universal bounds*. Note that defining Boolean algebra by Huntington's postulates is equivalent to defining it as a complemented and distributive lattice. In fact, it can be proved that a set $S$ in combination with two operations $(+)$ and $(\cdot)$ is a lattice iff the following laws are fulfilled:

*idempotency*: $a \cdot a = a + a = a$
*commutativity*: $a \cdot b = b \cdot a$ and $a + b = b + a$
*absorption*: $a + a \cdot b = a$ and $a \cdot (a + b) = a$
*associativity*: $a \cdot (b \cdot c) = (a \cdot b) \cdot c$ and $a + (b + c) = (a + b) + c$
*universal bounds*: $a + 0 = a$ and $a \cdot 0 = 0$ and $a \cdot 1 = a$ and $a + 1 = 1$

Boolean algebra is the fundamental basis for the analysis of digital electronic circuits. In the 1930s, Shannon [Shan38] showed that the behavior of switching circuits can be described by a two-valued Boolean algebra. Such a *switching algebra* is obtained if we consider a lattice defined by the set $B = \{0, 1\}$ and the operations of *conjunction* (*AND*), *disjunction* (*OR*) and *complementation* (*NOT*). The operation of conjunction is also called *product*, disjunction can also be referred to as *sum* and complementation is often called *negation*. These operations are defined in Table 1.1.

The switching algebra defined by the operations in Table 1.1 is isomorphic to a two-valued Boolean algebra given by Definition 1.2. Therefore, in the sequel, we will use the terms Boolean algebra and switching algebra interchangeably.

The properties of Boolean algebra are quite different from the conventional algebra of real numbers. In particular, idempotency, distributivity with respect to both operations $(+)$ and $(\cdot)$, and the existence of a complement mark special characteristics of Boolean algebra. Therefore, algorithms for logic synthesis, as will be discussed in Chapter 5, sometimes impose restrictions on the manipulations being performed on a switching circuit, excluding the above specialities of Boolean algebra. This reduces the algorithmic complexity and the resulting manipulations become analogous to manipulations in the conventional algebra of real numbers. Restrictions of this kind lead to the so-called *algebraic* techniques described in Section 5.3.

| $a$ | $b$ | $a \cdot b$ |
|---|---|---|
| 0 | 0 | 0 |
| 0 | 1 | 0 |
| 1 | 0 | 0 |
| 1 | 1 | 1 |

a) conjunction

| $a$ | $b$ | $a + b$ |
|---|---|---|
| 0 | 0 | 0 |
| 0 | 1 | 1 |
| 1 | 0 | 1 |
| 1 | 1 | 1 |

b) disjunction

| $a$ | $\bar{a}$ |
|---|---|
| 0 | 1 |
| 1 | 0 |

c) complementation

**Table 1.1:** Definitions of conjunction (AND), disjunction (OR) and complementation (NOT)

## 1.2 Graphs

A *graph* $G(V, E)$ is a pair $(V, E)$. $V$ is a set and $E$ is an incidence mapping on $V$. The elements of $V$ are called *vertices* or *nodes* of the graph and the elements of $E$ are called *edges*. Each edge $e \in E$ is associated with two nodes $v_i$, $v_j \in V$ being connected by this edge. If the nodes belonging to an edge are ordered the graph is called *directed*, otherwise it is called *undirected*. In the following we concentrate on directed graphs. In a directed graph, the node $v_i$ of an edge $(v_i, v_j)$ is the *immediate predecessor* of $v_j$, the node $v_j$ is called *immediate successor* of $v_i$.

A graph $G' = (V', E')$ is called a *subgraph* of $G = (V, E)$ iff $V' \subseteq V$ and $E' \subseteq E$. The number of edges incident to a node $v_i$ is called the *degree* of $v_i$. In a directed graph, the *outdegree* or *fanout* of a node refers to the number of its immediate successors. Similarly, the *indegree* or *fanin* is given by the number of its immediate predecessors. An alternating sequence of vertices and directed edges, which does not traverse any vertex more than once is called a *directed path* in the directed graph. If the first and the last nodes of a path are identical it is called a *cycle*. A directed graph without cycles is called a *directed acyclic graph* (*DAG*). DAGs represent partially ordered sets. If a DAG has one distinguished node, called *root*, which does not have any predecessors it is called *rooted*. A *directed rooted tree* is a DAG such that all nodes other than the root have exactly one immediate predecessor. The nodes without successors are called *leaves* of the tree. For simplicity, in this book we will refer to directed rooted trees simply as *trees*.

In a DAG, a node $v_j$ is called *successor* of a node $v_i$ and $v_i$ is called a *predecessor* of $v_j$ iff there is a directed path from $v_i$ to $v_j$. The set of all successors for a given node, in some literature, is also referred to as the *transitive fanout* of the node. Similarly, the predecessors of a node are called its *transitive fanin*.

Directed acyclic graphs are often used to describe structural properties of switching circuits. This is important for many heuristics as shown, e.g., for test generation in Section 2.5. Furthermore, DAGs can also serve as the basis for functional representations of switching circuits. The functional representations of Sections 1.3.4 and 4.2 associate Boolean functions with DAGs.

## 1.3 Boolean Functions and their Representations

Consider a Boolean algebra on a set $B$ and a set of variables $x_1, x_2, \ldots x_n$ such that each can be assigned independently an element of $B$. We say that the $n$ variables $x_1, x_2, \ldots x_n$ form an $n$-dimensional Boolean space $B^n$ where each vertex in the space is defined by one of the $|B|^n$ combinations of assignments. A *mapping*, $f: B^n \rightarrow B$ which uniquely associates every point of $B^n$ with an element of $B$ is called a Boolean *function*. In most applications it is $B = \{0, 1\}$. The vertices of $B^n$ being mapped to 1 are called the *on-set* of $f$ and those being mapped to 0 are called the *off-set* of $f$. Often, Boolean functions are incompletely specified, i.e., for some vertices in $B^n$ we "do not care" to what element of $B$ they are mapped. These vertices form the so called *don't care-set* of $f$. This is usually denoted by introducing a third element for $B$, denoted X, so that $B = \{0, 1, X\}$. Further, we speak of an $m$-ary Boolean function if each point in the space is mapped to $m$ elements of $B$, i.e., we consider a mapping $f: B^n \rightarrow B^m$.

Boolean functions are the basis to describe the behavior of switching circuits. Any method to optimize or analyze switching circuits relies on manipulating Boolean functions. Therefore, the efficiency of algorithms for specific problems in computer-aided design (CAD) of circuits depends highly on an appropriate *representation* of Boolean functions. In fact, the algorithmic solution of a given problem and the representation of the involved Boolean functions are intimately related and cannot be considered independent of each other. For simplicity, we speak of a *Boolean representation* when we mean the representation of a switching circuit as Boolean function.

Most commonly, Boolean functions are represented by Boolean *expressions* or sets of Boolean expressions. General Boolean expressions, as opposed to sum-of-product forms (to be defined) are also called *factored forms.*

**Definition 1.3:** A Boolean expression with the operations of disjunction, conjunction and negation is recursively defined by:

i) a variable is a Boolean expression.
ii) the constants 0 and 1 are Boolean expressions.
iii) the complement of a Boolean expression is a Boolean expression.
iv) the disjunction of two Boolean expressions is a Boolean expression.
v) the conjunction of two Boolean expressions is a Boolean expression.

A variable of a Boolean function in complemented or uncomplemented form is called a *literal*. For example, $a$ and $\bar{a}$ are expressions containing the same variable, however, they are two different literals.

### 1.3.1 Implementing Boolean Functions as Switching Circuit

Switching circuits are constructed by interconnections of electronic *gates* which implement elementary Boolean functions like disjunction, conjunction or complementation. Figure 1.1 depicts some standard logic gate types. Gates are electronic devices that produce voltage levels at their outputs as a function of the voltage levels received at the inputs. These voltage levels are restricted to two ranges "high" and "low" and are usually associated with the logic values 1 and 0. Boolean functions can be implemented by an interconnection of electronic gates. If *all* Boolean functions can be implemented using a given set of gates then the set of gates is called *functionally complete*. For example {AND, OR, NOT} is functionally complete. Actually, either AND or OR can be removed from this set without making it incomplete, i.e., either {NAND} or {NOR}, each represents a functionally complete set of gates. Boolean functions can also be expressed exclusively by AND and XOR if the logic value 1 is available as constant input for the logic gates. Therefore, {1, AND, XOR} is a functionally complete set.

The implementation of a Boolean function by a switching circuit can be derived directly from the mathematical representation of the function. However, not every Boolean representation is equally suitable as a basis for a low cost implementation.

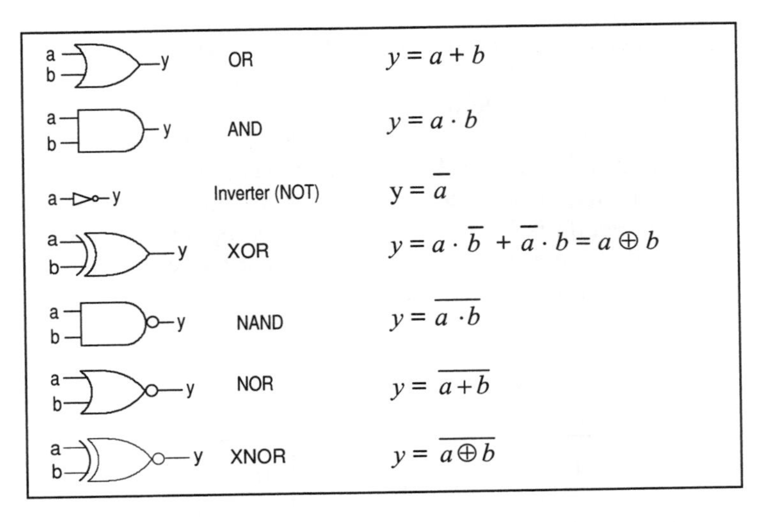


**Figure 1.1:** Standard electronic gates and their symbols

### 1.3.2 Disjunctive Forms

Let $B$ denote a set forming a Boolean algebra. In some literature, the set $B$ is referred to as *logic alphabet* and its elements are called *logic values*. The most naive way to represent a Boolean function is to list all vertices of the Boolean space as rows in a table and to associate with each row a logic value. An example of such a *truth table* is shown in Figure 1.2.

| $x_1$ | $x_2$ | $x_3$ | $y$ |
|---|---|---|---|
| 0 | 0 | 0 | 0 |
| 0 | 0 | 1 | 1 |
| 0 | 1 | 0 | 0 |
| 0 | 1 | 1 | 1 |
| 1 | 0 | 0 | 0 |
| 1 | 0 | 1 | 0 |
| 1 | 1 | 0 | 1 |
| 1 | 1 | 1 | 0 |

**Figure 1.2:** Example of a tabular Boolean representation (truth table)

Since the truth table contains $2^n$ rows for a function with $n$ variables, Boolean representation by truth tables is not practical except for very small examples.

The following notions are useful to describe other representations of Boolean functions: a conjunction of literals is called a *product term*. A product term that contains a literal for every variable of a Boolean function $f$ is called *minterm* of $f$. Similarly, a *sum term* is a disjunction of literals and if it contains a literal for every variable in $f$ it is called a *maxterm*. Minterms and maxterms can be associated with the rows of a truth table. If a minterm is associated with a row where $f$ is 1 it is called *on-set minterm*, if $f$ is 0 it is called *off-set minterm*. Similarly, we can distinguish *on-set maxterms* and *off-set maxterms*. For the above example, Figure 1.3 shows the on-set minterms and off-set maxterms.

| $x_1$ | $x_2$ | $x_3$ | $f$ | *terms* |
|---|---|---|---|---|
| 0 | 0 | 0 | 0 | $M_1 = x_1 + x_2 + x_3$ |
| 0 | 0 | 1 | 1 | $m_1 = \overline{x_1} \cdot \overline{x_2} \cdot x_3$ |
| 0 | 1 | 0 | 0 | $M_2 = x_1 + \overline{x_2} + x_3$ |
| 0 | 1 | 1 | 1 | $m_2 = \overline{x_1} \cdot x_2 \cdot x_3$ |
| 1 | 0 | 0 | 0 | $M_3 = \overline{x_1} + x_2 + x_3$ |
| 1 | 0 | 1 | 0 | $M_4 = \overline{x_1} + x_2 + \overline{x_3}$ |
| 1 | 1 | 0 | 1 | $m_3 = x_1 \cdot x_2 \cdot \overline{x_3}$ |
| 1 | 1 | 1 | 0 | $M_5 = \overline{x_1} + \overline{x_2} + \overline{x_3}$ |

**Figure 1.3:** Minterms and maxterms for a Boolean function

A Boolean function is completely described by a disjunction of all its on-set minterms or by a conjunction of all its off-set maxterms. In the former case we speak of the *disjunctive normal form* (*DNF*) and in the latter we speak of the *conjunctive normal form* (*CNF*). In the example of Figure 1.3 the normal forms are:

$$\begin{aligned} y &= \overline{x_1} \cdot \overline{x_2} \cdot x_3 + \overline{x_1} \cdot x_2 \cdot x_3 + x_1 \cdot x_2 \cdot \overline{x_3} && \text{(DNF)} \\ &= (x_1 + x_2 + x_3) \cdot (x_1 + \overline{x_2} + x_3) \cdot (\overline{x_1} + x_2 + x_3) \cdot \\ &\quad (\overline{x_1} + x_2 + \overline{x_3}) \cdot (\overline{x_1} + \overline{x_2} + \overline{x_3}) && \text{(CNF)} \end{aligned}$$

These normal forms are *unique* representations of Boolean functions, i.e., two Boolean functions are equivalent if and only if their normal forms contain the same minterms or maxterms, respectively. Unique representations of Boolean functions are also called *canonical*. Although conjunctive and disjunctive forms are equally important it is usually sufficient to only refer to one of them when describing notions and algorithms of switching theory. The standard literature gives preference to the disjunctive form. Therefore, the following considerations will be restricted to the disjunctive form.

A general Boolean expression can be transformed into its DNF by repeatedly applying Shannon's expansion theorem, given by

$$y(x_1,\dots x_n) = x_i \cdot y(x_1,\dots x_i = 1,\dots x_n) + \overline{x_i} \cdot y(x_1,\dots x_i = 0,\dots x_n).$$

In short form we write:

$$y = x \cdot y|_{x=1} + \overline{x} \cdot y|_{x=0}$$

Shannon's expansion is a special case of an orthonormal expansion in terms of some orthogonal function (see [Brow90]). This will be further discussed in Section 5.6.1 where this is used to formulate a test generation based method to optimize switching circuits.

Further, the notion of an *implicant* is very important in the theory of optimizing switching functions. An *implicant* for a function $f$ is a product term $t$ such that $t = 1 \Rightarrow f = 1$. The implicant is called *prime* if the product term is not an implicant anymore if any literal is removed. A set of implicants is said to *cover* (to be a *cover* for) a function $f$ iff for every minterm $m$ of $f$, the set contains an implicant $t$ with $m = 1 \Rightarrow t = 1$. It is common to represent Boolean functions by a disjunction of implicants referred to as *sum of products* (*SOPs*). In contrast, a general Boolean expression according to Definition 1.3 which is not a SOP is often called a *factored form*.

It is often desirable to make a SOP as small as possible. If a SOP only contains prime implicants the SOP is called *prime*. If the function is not covered anymore if any of the implicants is removed, the SOP is called *irredundant*. Note that a prime and irredundant SOP is not a canonical representation of a Boolean function. There usually exist many possibilities to cover a Boolean function by a selection of prime implicants. Finding the right selection, i.e., determining a set of prime implicants that covers the function and leads to minimal cost when implementing the SOP as an electronic circuit, is the classical problem of two-level minimization. The first exact solution to this problem has been given by Quine [Quin52] and McCluskey [McCl56] and is known as the *Quine-McCluskey method*. For large circuits an exact solution may not be viable. Therefore, heuristic minimization techniques have been presented as, e.g., in ESPRESSO [Bray84]. These heuristic methods heavily rely on exploiting properties of *unate* functions or sub-functions. A SOP-expression is called *unate* if each variable appears only in its complemented or uncomplemented form, but not both. For example, $y = \bar{a}b + bc$ is unate but $y = ab + \bar{b}c$ is not. A Boolean function is called unate, iff there exists a unate SOP-expression for it.

### 1.3.3 Boolean Networks

A circuit implementation of the forms described in Section 1.3.2 results in *two-level* circuits, i.e., any path from the inputs to the outputs of the circuit traverses at most two gates. In general, smaller circuits can be obtained if no restrictions are made on the number of levels in the circuit. Such *multi-level* circuits are usually described by *Boolean networks*.

**Definition 1.4:** A *Boolean network* is a triple $(V, E, F)$, with the directed acyclic graph $(V, E)$ and a set of Boolean functions $F$.

i) The set $I \subseteq V$ is a set of nodes without predecessors which correspond to distinct nodes called *primary inputs* of the Boolean network. The set $O \subseteq V$ is the set of nodes without successors called *primary outputs*. The edges leaving the nodes of $I$ are associated with the input variables of the Boolean network.

ii) The nodes $g_i \in V \setminus I$ are associated with Boolean functions $f_i \in F$ and the edges correspond to variables $s_i$ denoting the functions $f_i$ of the nodes which they are leaving. For every function $f_j$ in the network there is an edge from $g_i$ to $g_j$ if $f_j$ depends on $s_i$.

Boolean networks are a technology-independent description of combinational circuits. Throughout this book we consider circuits being described by *gate netlists*. The gates are elements of some library $L$ which depends on the available technology and the specific application. A gate netlist description is a special case of a Boolean network where each gate is associated with a node and the function of the node is defined by the gate function. In this work we generally assume that the library $L$ consists only of the gate types shown in Figure 1.1. AND, OR, NOR, NAND can have an arbitrary number of inputs, and XOR, XNOR must have exactly two inputs. Note that this choice of $L$ corresponds to the usual assumptions in literature and contains the elements of a typical gate netlist description. Such a restricted Boolean network, in the sequel, will be referred to as *combinational network* or *combinational circuit* or *circuit netlist* interchangeably. Further, avoiding formalism we often denote a node, function and variable in a combinational network with the same symbol and speak of nodes, functions, variables or signals interchangeably.

### 1.3.4 Binary Decision Diagrams

*Binary Decision Diagrams* (*BDDs*) are graph representations of Boolean functions. The motivation to represent Boolean functions by *binary decision diagrams* is twofold: firstly, certain types of binary decision diagrams are *canonical* representations of Boolean functions. This is important in many applications, especially in formal verification. Secondly, certain algorithmic problems and manipulations of Boolean functions which have exponential worst case complexity for conventional representations as described in Sections 1.3.2 or 1.3.3, only have polynomial complexity for BDDs. In particular, this is true for the important *satisfiability* problem for a Boolean function, i.e., the problem of determining whether or not a Boolean function can evaluate to the logic value 1. This problem is *NP-complete* if the function is represented as a product of sums (conjunctive form), it is a constant time operation if the function is represented by a BDD.

Binary decision diagrams are special types of *branching programs*. For more information on branching programs see, e.g., [Wege87]. It was first suggested by Akers [Aker78] to use binary decision diagrams as a Boolean representation for solving problems in design automation and particularly in test generation. This became practical by a refinement of the model and the introduction of *reduced ordered binary decision diagrams* (*ROBDDs*) by Bryant [Brya86]. Bryant also developed the basic algorithms for Boolean manipulations using ROBDDs.

Ordered binary decision diagrams (OBDDs) as proposed by Bryant [Brya86] can be defined as follows:

> **Definition 1.5:** An *OBDD* is a rooted DAG with vertex set $V$. Each non-leaf vertex has as attributes a pointer $index(v) \in \{1, 2, \ldots, n\}$ to an input variable in the set $\{x_1, x_2, \ldots, x_n\}$, and two children $low(v), high(v) \in V$. A leaf vertex $v$ has as an attribute a value $value(v) \in B$.
>
> For any vertex pair $\{v, c_v\}$, $c_v \in \{low(v), high(v)\}$, such that no vertex is a leaf, $index(v) < index(c_v)$.

Roughly speaking, an OBDD represents a Boolean function if we associate the steps of a Shannon expansion with the nodes of the OBDD. This can be defined more precisely as follows:

**Definition 1.6:** An OBDD with root $v$ denotes a function $f^v$ such that:

i) if $v$ is a leaf with *value*($v$) = 1, then $f^v = 1$.
ii) if $v$ is a leaf with *value*($v$) = 0, then $f^v = 0$.
iii) if $v$ is not a leaf and *index*($v$) = $i$, then $f^v = \overline{x_i} \cdot f^{low(v)} + x_i \cdot f^{high(v)}$.

**Example 1.1:** As an example we construct an OBDD for function $f = a \cdot b + c$. Figure 1.4 shows the corresponding OBDD. As a convention, the child *low*($v$) is always attached to the edge leaving $v$ towards the left. The child *high*($v$) is attached to the edge on the right. We build the OBDD by repeatedly performing Shannon's expansion. This illustrates the construction rule of OBDDs as given in Definition 1.6. It should be mentioned however that this is *not* quite how OBDDs are actually constructed for a given circuit description in practical applications.

The root node can be associated with function $f$. Performing a Shannon expansion for variable $a$ we obtain the two cofactors $c$ and $b + c$. These correspond to the left and right children of the root node. We now decompose these cofactors by further applications of Shannon's expansion and this is continued until constant values 0 and 1 are obtained as cofactors. In an *ordered* BDD the variables are always picked in the same order. For function $f$ we have assumed a variable order ($a$, $b$, $c$), i.e., according to Definition 1.5 we choose the indices as: *index*($a$) = 1, *index*($b$) = 2, *index*($c$) = 3.

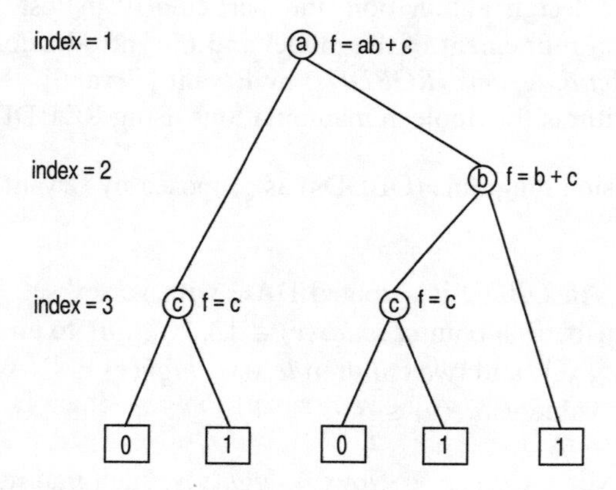


**Figure 1.4:** OBDD for function $f = a \cdot b + c$

The OBDD in Figure 1.4 can still be *reduced*. Only a reduced OBDD (ROBDD) is canonical. An OBDD is reduced by checking whether the graph contains isomorphic subgraphs.

**Definition 1.7:** Two OBDDs $F = (V, E)$ and $F' = (V', E')$ are called *isomorphic* if and only if there exists a bijective function $m: V \rightarrow V'$ such that for all $v \in V$, $m(v) \in V'$:

i) $v$ and $m(v)$ are leaf vertices with the same value,
$value(v) = value(m(v))$, or

ii) $v$ and $m(v)$ are non-leaf vertices with the same index,
$index(v) = index(m(v))$.
Further it is $m(low(v)) = low(m(v))$ and $m(high(v)) = high(m(v))$.

In other words, two OBDDs are isomorphic if there is a one-to-one mapping between all nodes of the same type such that the children of a node in one OBDD are mapped onto the children of the corresponding node in the other OBDD.

**Example 1.1 (contd.):** Consider the OBDD in Figure 1.4. It contains isomorphic subgraphs, namely the subgraphs rooted in node *c*. By sharing these isomorphic subgraphs we obtain the reduced OBDD shown in Figure 1.5.

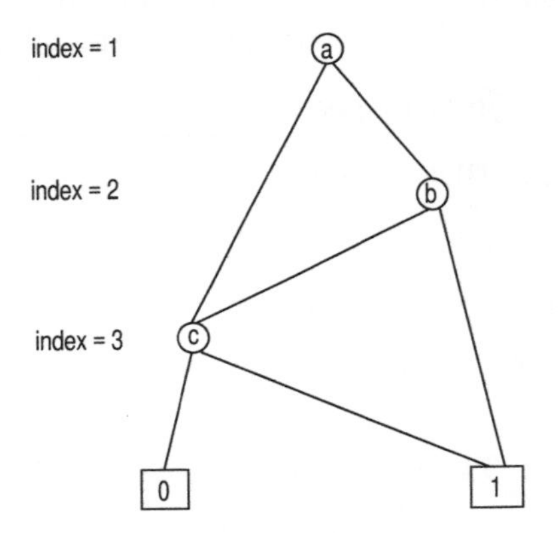


**Figure 1.5:** Reduced OBDD of Figure 1.4

**Definition 1.8:** An OBDD is called *reduced* OBDD (ROBDD) if it contains no vertex $v$ with $low(v) = high(v)$, nor any nodes $v, v' \in V$ such that the subgraphs rooted in $v$ and $v'$ are isomorphic.

In an OBDD, every path from the root to one of the leaves corresponds to a combination of value assignments for the variables of the represented function. If the path leaves a node through the left edge this means assigning 0 to the corresponding variable, the right edge corresponds to assigning a 1. If the path terminates in a leaf with value 1 then the set of value assignments makes the function evaluate to 1, otherwise, for the leaf with value 0 the function is 0. If the function is not satisfiable then there only exists a leaf with value 0 and the ROBDD consists only of this node.

ROBDDs are an important Boolean representation facilitating the solution of many problems in circuit design. However, in order to benefit from this representation it is essential to develop efficient algorithms to perform operations such as

- *apply*: given two OBDDs for functions $f_1$ and $f_2$, determine an OBDD for the function that results, if a two-input Boolean function like conjunction (AND), disjunction (OR), antivalence (XOR), etc., is applied to $f_1$ and $f_2$.
- *compose*: given two OBDDs for functions $f_1$ and $f_2$, determine an OBDD for function $f_1$ if one variable $x_i$ of $f_1$ is replaced by function $f_2$.

Table 1.2 shows the complexity of some basic operations on binary decision diagrams as proposed by Bryant [Brya86]. For a function $f$ its OBDD is denoted by $F$ and the number of vertices in $F$ is $|F|$. The implementation of these operations is based on the *if-then-else-operator* (*ITE-operator*) introduced by Brace [Brac90]. All operations in Table 1.2 can also be expressed in terms of the ITE-operator. For a more detailed description of these operations see [Brac90].

| OPERATION | | COMPLEXITY |
|---|---|---|
| *reduce:* | makes OBDD canonical | $O(\|F\|)$ |
| *apply:* | $y = y(f_1, f_2)$ | $O(\|F_1\| \cdot \|F_2\|)$ |
| *restrict:* | $f_1 = (x_1, \ldots x_i = V \ldots x_n),\ V \in \{0, 1\}$ | $O(\|F\|)$ |
| *compose:* | $f_1 = (x_1, \ldots x_i = f_2 \ldots x_n)$ | $O(\|F_1\|^2 \cdot \|F_2\|)$ |

**Table 1.2:** Operations on OBDDs

OBDDs are usually built using the above operations. If an OBDD is to be built for a multi-level combinational network this can be accomplished using the *apply* operation. Starting at the primary inputs of the circuit as a first step OBDDs are built for the circuit nodes adjacent to the primary inputs. Moving towards the primary outputs using the *apply* operation, OBDDs are built step by step for the internal circuit nodes until the primary outputs are reached. *Hashing* plays an important role in this procedure aiming at building new OBDDs making maximum use out of previously computed OBDDs. Alternatively, the *compose* operation could be used to build an OBDD for a multi-level circuit. In this case, we have to move backwards from the primary outputs to inputs. Since *compose* is more complex than *apply*, as shown in Table 1.2, the *apply* operation generally would be preferred.

*Variable Ordering* is important in making OBDD-based methods efficient. For many practical functions the ROBDD size is highly sensitive to the variable ordering. Therefore, many heuristics have been developed to determine and change the variable orderings for OBDDs [MaWa88], [FuMa91], [IsSa91], [Rude93]. Unfortunately, for some functions like integer multiplication the size of the ROBDD is exponential in the number of input variables no matter what variable ordering is chosen [Brya86]. Therefore, some circuits like multipliers are not amenable to BDD-based techniques.

The fact that ordered binary decision diagrams are canonical representations of Boolean functions implies that the *structural* information about a given circuit implementation is completely lost when the circuit is represented as OBDD. This leads to advantages when performing the described manipulations or when solving certain problems like those encountered in formal verification. On the other hand, it is often of great advantage to preserve some structural information, especially when problems related to logic synthesis are considered. This motivates the work described in Chapter 4 which presents a more structure-oriented Boolean representation based on *AND/OR reasoning graphs*. Methods based on AND/OR reasoning graphs follow a very different paradigm when compared to BDDs. It will be shown in Chapter 5 and 6 that this is very beneficial for certain applications and it can be expected that an appropriate mix of functional techniques based on BDDs and structural techniques such as the ones to be described will constitute adequate solutions to many problems.

In this context it should be noted that concepts have been developed to incorporate structural information directly into a BDD-type representation. In 1976, Ubar described *alternative graphs* [Ubar76] which are general binary decision diagrams on which a certain superposition procedure is applied to incorporate structural information. As a result the nodes in the alternative graph correspond to certain paths in the circuit. This is used to simplify the testing problem in combinational circuits. For more information refer to [Ubar94].

# Chapter 2

# COMBINATIONAL ATPG

*Automatic Test Pattern Generation* (*ATPG*) for combinational circuits has been an active field of research for many years. This chapter introduces the basic algorithmic concepts of deterministic test generation for single stuck-at faults in combinational circuits. Today, combinational ATPG for single stuck-at faults is considered quite mature. Most modern tools are based on the contributions of [Roth66], [Goel81], [FuSh83], [ScTr87] and [Larr89].

Recently, it has been shown that basic algorithmic concepts of combinational ATPG are also useful in solving other problems of switching theory like *logic verification* [Bran93], [Kunz93] and *logic optimization* [EnCh93], [KuMe94]. This has led to renewed interest in combinational ATPG. Therefore, in this chapter, we present the basic notions of common ATPG algorithms and give special attention to concepts that are important for a wider range of applications. Specifically, we focus on the so called *structural* ATPG methods as they have been derived from Roth's D-algorithm [Roth66]. The application of ATPG to other CAD problems will be the subject of Chapters 5 and 6.

## 2.1 Problem Formulation

The goal of testing is to detect physical defects on a chip inflicted by the fabrication process or those occurring later during the operation of the chip. In this book, the physical aspects of testing are completely ignored. All considerations are restricted to combinational circuits at the abstraction of the logic gate-level, i.e., the circuit is represented as a *combinational network*, defined in Section 1.3.3. In this book, our interest is exclusively in the algorithmic concepts of test generation, especially those that can also be applied to other design automation problems.

Physical defects are described on the logic gate level description by certain *fault models*. The *single stuck-at fault* is a simple model and assumes that a *single* line in the combinational circuit fails to change its logic value and is "stuck" at a constant value of 0 or 1. This fault model has become widely accepted since practical and theoretical investigations have shown that a set of tests detecting all single stuck-at faults in a circuit usually detects the majority of practically relevant physical defects including those that cannot be modeled in this simple way.

Techniques of test generation can be roughly divided into *random* and *deterministic* methods. In random methods, a set of *random* input stimuli is generated for the *Circuit Under Test* (CUT). A *fault simulator* can be used to measure the quality of the random stimuli with respect to detecting a given set of faults. For a survey of fault simulation techniques, see e.g. [AbBr90]. Random techniques are particularly of interest in the context of *design for testability*, see e.g. [McCl86], [AbBr90].

In *deterministic* ATPG, the starting point of all operations is a list of faults that must be detected by the tests to be determined. The faults in the fault list are targeted one after other and for every fault a test is calculated by the ATPG algorithm.

In this chapter we exclusively consider *deterministic* ATPG. For reasons of simplicity the word "deterministic" is often omitted. The following problem formulation simplifies the general deterministic test generation problem by two assumptions:

i) The circuit is *combinational* and only contains primitive gates of types: AND, OR, NOT, NAND, NOR, XOR, XNOR; i.e., it is represented as a *combinational network*, defined in Section 1.3.3.

ii) Only single stuck-at faults are considered.

The problem of deterministic test generation is illustrated in Figure 2.1. As mentioned, deterministic test generation is always fault oriented, i.e., for a given circuit $C$ there is a fault list $\Phi = \{\phi_1, \phi_2, \ldots \phi_k\}$ containing all faults for which a test has to be generated. These faults are called *target faults*. Let circuit $C$ have $n$ primary input signals and $m$ primary output signals and let $\boldsymbol{y} = (y_1 = V_1, y_2 = V_2, \ldots y_m = V_m)$ be the response of the *fault-free* circuit to some input stimulus $\boldsymbol{x} = \{x_1, = W_1, x_2 = W_2, \ldots x_n = W_n\}$ with $V_i, W_j \in \{0, 1\}$. The input stimulus $\boldsymbol{x}$ is called a *test* for a fault $\phi$ if and only if the circuit response $\boldsymbol{w}$ in the presence of $\phi$ is different from the response $\boldsymbol{y}$ of the fault-free circuit, $\boldsymbol{w} \neq \boldsymbol{y}$. Not for every fault in a circuit there exists a test. Untestable faults are also called *redundant*. The task of a deterministic ATPG-algorithm is to calculate a test $\boldsymbol{x}$ for a given fault $\phi$ if a test exists, or to prove its untestability otherwise.

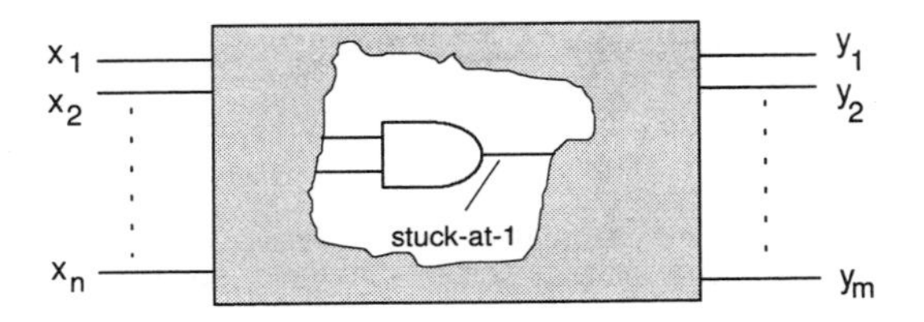


**Figure 2.1:** Circuit under test (CUT)

Goel [Goel81] has formulated deterministic test generation for a given fault $\phi$ as a search in the finite Boolean space $B = \{0, 1\}^n$. The Boolean space is explored systematically until an input stimulus $\boldsymbol{x}$ has been found that represents a test. Therefore, the input stimuli $\boldsymbol{x}$ and circuit responses $\boldsymbol{y}$ are often referred to as input *vectors* and output *vectors*. An input vector being able to detect a fault $\phi$ is called test *vector* or test *pattern* for fault $\phi$.

If an input stimulus $\boldsymbol{x}$ is a test vector for a fault $\phi$ we say that $\phi$ is *covered* by $\boldsymbol{x}$. The task of a deterministic test generator is to determine a set of tests for a given circuit $C$ and a fault list $\Phi$, such that every testable fault in $\Phi$ is covered. Sometimes the test generation process for specific faults may have to be aborted because the computational costs turn out to be too high. An important measure for the quality of a test set generated by some (random or deterministic) test generation method is the *fault coverage* defined by

$$\textit{fault coverage} = \frac{\text{number of faults covered by test set}}{\text{number of faults in fault list}} \cdot 100\ \%.$$

In order to evaluate the performance of a test generator it is common to assume that only aborted faults are not covered by the generated vectors. This leads to a quality measure which is often called *fault efficiency*. It is calculated in the same way as fault coverage except that the number of identified redundancies is subtracted from the number of faults in the fault list.

Algorithms for deterministically generating a test vector for a given fault $\phi$ are usually embedded in a general test generation procedure as shown in Figure 2.2. Fault simulation is often used to accelerate test generation by reducing the list of faults before and during the ATPG process. Figure 2.2 shows the flowchart of a frequently used procedure to derive a deterministic test set for a given circuit and a given set of faults.

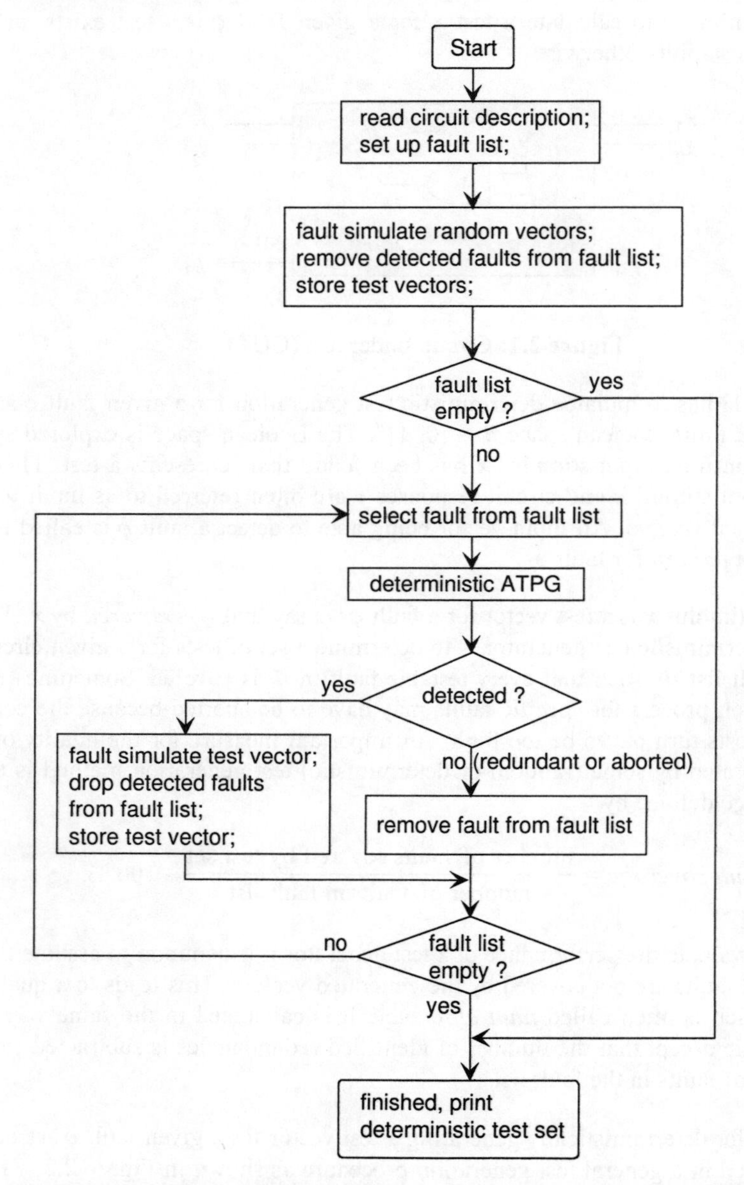


**Figure 2.2:** Test generation procedure with simulation and fault dropping

As a first step, it is beneficial to run fault simulation using randomly generated test vectors. The number of vectors usually ranges between a few hundred and several thousand. For each random vector that detects a target fault in the fault list, this fault is removed from the fault list and the test vector is stored. This random phase can shrink the fault list drastically, often leaving only few faults as targets for the deterministic ATPG-algorithm. Next the deterministic ATPG-algorithm targets one of the remaining faults in the fault list and attempts to calculate a test vector. If a test vector is calculated it is stored. Often a calculated test vector may detect more faults in the fault list than just the considered target fault. Therefore, a fault simulator is used to simulate the calculated test vector in order to *drop* all faults from the fault list that are also detected by this test vector. This is called *fault dropping*. The procedure is repeated until the fault list is empty. The stored test vectors represent a deterministic test set for the CUT.

Ibarra and Sahni [IbSa75] have shown that the problem of determining whether or not a single stuck-at fault is testable belongs to the class of *NP-complete* problems. This leaves little hope that a test generator will ever be developed performing well for all classes of circuits. In the *worst case* the computational complexity of test generation grows exponentially with the size of the circuit. Therefore, it is of crucial importance to guide the search for test vectors such that the worst case is avoided as often as possible. This is the task of *heuristics*. Modern ATPG-tools typically make use of a large variety of efficient heuristics that have been reported over the years so that the theoretical worst case can be avoided in most practical cases. In the following sections, along with basic algorithmic concepts, the most important heuristics for deterministic ATPG in combinational circuits will be described.

## 2.2 The Decision Tree

Generally, along the search for a solution to a specific problem *decisions* have to be made. Also test generation can be viewed as a search process consisting of a sequence of decisions. Unfortunately, decisions can be *wrong*. At a given search state, a decision is wrong if no solution exists for all remaining possibilities of future decisions. It is therefore unavoidable to reverse such a decision in order to find a solution. A classical scheme to keep track of the decisions made and examine all possible alternatives is the *decision tree*.

The decision tree is a tree, each node representing a problem to be solved by the algorithm. Each edge leaving a node represents a decision being possible at this node. Figure 2.3 shows an example of a decision tree.

A decision tree is generally traversed *depth-first* following a *branch-and-bound* strategy. This can be illustrated with Figure 2.3. The starting point is a given problem $A$ allowing for two decisions $A1$ and $A2$. If we make decision $A1$ this leads to problem $B$ (*branching*). Also at problem $B$ there exist several possibilities and we choose to continue the search with decision $B1$ (*branching*) so that problem $C$ arises. At this point, it turns out that both possible decisions at problem $C$ lead to a *conflict*, i.e., we recognize that neither for decision $C1$ nor decision $C2$ a solution can be found. Since there are no other alternatives at problem $C$, it is mandatory to reverse decision $B1$, i.e., the node associated with problem $C$ must be pruned from the decision tree (*bounding*). Now decision $B2$ is taken for problem $B$ (*branching*). Since this also leads to a conflict we are forced to take decision $B3$. This scheme is continued until either a solution has been found or it is certain that no solution exists. The latter case occurs if the root node has been pruned from the decision tree.

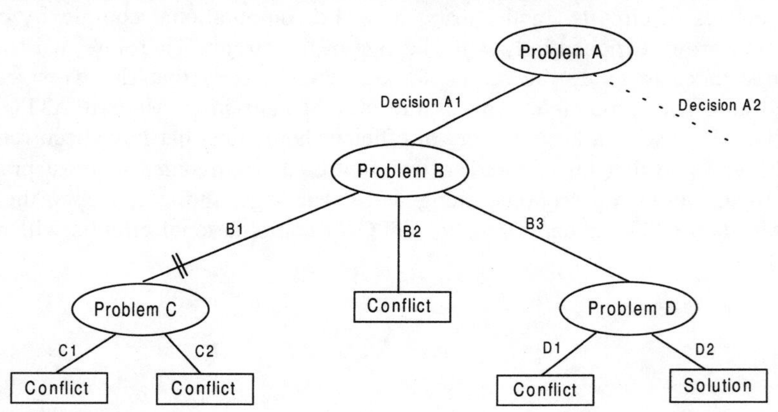


**Figure 2.3:** Decision tree

Most common test generators rely on a decision tree and follow the above scheme as proposed by Goel [Goel81]. In test generation, each node in the decision tree corresponds to a signal in the circuit being subject to a decision. Initially, all signals in the circuit are assigned X (unknown). A decision is made by assigning a binary logic value, 0 or 1, to a selected signal. Each branch in the decision tree is associated with the decision (value assignment) being taken at this node (signal). If a sequence of decisions has led to a conflict, e.g., there is a logic inconsistency between signal values in the circuit, then alternative value assignments have to be explored as illustrated above. This process is also referred to as *backtracking*. If the decision tree is completely exhausted, it is guaranteed that all combinations of value assignments at the considered signals are taken into account *implicitly*, without *explicitly* making all possible combinations of binary value assignments. This is possible because some sub-trees of the decision tree can often be recognized as non-solution areas and need not be explored. Therefore, the search based on a decision tree in some literature is also called *implicit enumeration* as opposed to *explicit enumeration.* See [ChBu96] for a detailed treatment of branch-and-bound methods in test generation and design automation.

Note the following important property of the decision tree: at any moment during the search a single path through the graph is sufficient to completely describe the current state of the search process. Therefore, the decision tree can simply be implemented as a *stack* where every element in the stack corresponds to a node in the decision tree and keeps track of decisions already made at this node. If there are only two possible decisions at every node, i.e., if the decision tree is binary as in most test generation techniques, this information can simply be stored in a single *flag* belonging to every element in the stack. Branching and bounding in the decision tree then simply correspond to *pushing* and *popping* of the stack. Memory requirements for implicit enumeration are therefore not determined by the number of decisions to be made, but only depend linearly on the number of nodes at which there are valid decisions at a given instance of time.

## 2.3 Logic Value Alphabets and Implications

A combinational circuit with $n$ primary inputs and $m$ primary outputs is specified on the logic level by an $m$-ary switching function $f$: $B^n \rightarrow B^m$, where $B$ is a domain of logic values as described in Section 1.3. For fault-free circuits it is common to use the logic alphabet $B_2 = \{0, 1\}$. Often, for specific problems other logic alphabets are required. In order to describe the faulty behavior of a circuit Roth's *D-calculus* [Roth66] has become widely accepted. In Roth's notation a signal is assigned the logic value D if it assumes 1 in the fault-free and 0 in the faulty circuit. In the opposite case, if the signal is 0 in the fault-free and 1 in the faulty circuit, it is denoted by $\overline{\text{D}}$. For logic values being equal in the fault-free and faulty cases, namely 0 or 1, the signal value is denoted 0 or 1, respectively. With these notations we obtain the logic alphabet $B_4 = \{0, 1, \text{D}, \overline{\text{D}}\}$ and it can be verified that also $B_4$ together with the operations of disjunction, conjunction and negation as defined in Section 1.3 forms a Boolean algebra. For test generation it is of advantage to introduce a fifth logic value, X, describing the case where no unique logic value has been assigned. This value is usually referred to as the *don't care* or *unknown* value. Including X into $B_4$ results in the five-valued logic alphabet $B_5 = \{0, 1, X, D, \overline{D}\}$ which to this date forms the basis for many modern ATPG-algorithms. With the above definitions of logic values 0, 1, X, D, $\overline{D}$ and the operations of conjunction, disjunction and negation defined in Table 1.1 we obtain the truth tables for this five-valued logic as shown in Table 2.1.

It is interesting to observe that $B_5$ and the operations of Table 2.1 do not form a Boolean algebra. Looking at the truth tables of Table 2.1 it can be noted that the *associative law* is violated by the five-valued logic alphabet. We obtain:

$$D \cdot (\overline{D} \cdot X) = D \cdot X = X$$
$$(D \cdot \overline{D}) \cdot X = 0 \cdot X = 0$$

The reason for this problem lies in the insufficient resolution of the don't care value. For example, a signal is assigned the value X if any of the logic values {0, 1, D, $\overline{D}$} is a possible future assignment. Note that it is also assigned X in a different situation where only the values {0, D} can occur. To solve this problem Akers suggested a 16-valued logic alphabet [Aker76] obtained by the power set $\mathcal{P}(B_4)$, with $B_4 = \{0, 1, D, \overline{D}\}$. This logic alphabet is shown in Table 2.2.

| AND | 0 | 1 | X | D | $\overline{D}$ |
|---|---|---|---|---|---|
| 0 | 0 | 0 | 0 | 0 | 0 |
| 1 | 0 | 1 | X | D | $\overline{D}$ |
| X | 0 | X | X | X | X |
| D | 0 | D | X | D | 0 |
| $\overline{D}$ | 0 | $\overline{D}$ | X | 0 | $\overline{D}$ |

| OR | 0 | 1 | X | D | $\overline{D}$ |
|---|---|---|---|---|---|
| 0 | 0 | 1 | X | D | $\overline{D}$ |
| 1 | 1 | 1 | 1 | 1 | 1 |
| X | X | 1 | X | X | X |
| D | D | 1 | X | D | 1 |
| $\overline{D}$ | $\overline{D}$ | 1 | X | 1 | $\overline{D}$ |

| $f$ | 0 | 1 | X | D | $\overline{D}$ |
|---|---|---|---|---|---|
| NOT($f$) | 1 | 0 | X | $\overline{D}$ | D |

**Table 2.1:** AND-, OR-, NOT-operation in 5-valued logic (*D-calculus*)

It can be shown that Akers' 16-valued logic alphabet forms a Boolean algebra. A test generator based on 16-valued logic has been presented by Rajski and Cox [RaCo90]. By choosing subsets of $B_{16}$ it is possible to create further logic alphabets. Particularly, some test generators [KiMe87], [MaGr90] use a 9-valued logic system which was originally proposed by Muth [Muth76].

The choice of an appropriate logic alphabet has notable influence on the performance of a test generator. More logic values permit better resolution of the signal values and therefore provide a more complete picture on the current situation of value assignments in the circuit. On the other hand, a larger number of logic values causes administrative overhead, which is not always compensated by the additional information gained. For test generation for single stuck-at faults in combinational circuits, the above 5-valued logic system, in spite of its insufficiencies, has encountered wide-spread popularity. Throughout the rest of this book, depending on the context, we will only consider the logic alphabets $B_2 = \{0, 1\}$, $B_3 = \{0, 1, X\}$, and $B_5 = \{0, 1, X, D, \overline{D}\}$. The methods developed, in principle, can also be applied to other logic alphabets like Akers' $B_{16}$ but to avoid formalism they are formulated only for these common alphabets.

| $B_{16}$ | $\mathcal{P}(B_4)$ |
|---|---|
| 0 | $\{\}$ |
| 1 | $\{0\}$ |
| 2 | $\{\overline{D}\}$ |
| 3 | $\{0, \overline{D}\}$ |
| 4 | $\{D\}$ |
| 5 | $\{0, D\}$ |
| 6 | $\{\overline{D}, D\}$ |
| 7 | $\{0, \overline{D}, D\}$ |
| 8 | $\{1\}$ |
| 9 | $\{0, 1\}$ |
| 10 | $\{\overline{D}, 1\}$ |
| 11 | $\{0, \overline{D}, 1\}$ |
| 12 | $\{D, 1\}$ |
| 13 | $\{0, D, 1\}$ |
| 14 | $\{\overline{D}, D, 1\}$ |
| 15 | $\{0, \overline{D}, D, 1\}$ |

**Table 2.2:** 16-valued logic [Aker76]

The following terminology is used:

> **Definition 2.1:** For the logic alphabets $B_2$, $B_3$, $B_5$ the values 0, 1, D and $\bar{D}$ are called *fixed* or *specified*. The logic value X is called *unspecified*.

Besides the logic alphabet, the *implication procedure* is of great importance when a given situation of value assignments is to be evaluated. In the context of test generation, the process of determining value assignments being necessary for the consistency of a given set of value assignments is referred to as *performing implications*. Like in [KuPr93] the following classification of implications is used:

> **Definition 2.2:** In a combinational network, a value assignment at an arbitrary input or output signal of a logic gate $g$ follows by *simple implication* if it is uniquely determined by previous logic value assignments at other input or output signals of the same gate $g$ and by the function of $g$ (AND, OR, NOT, NAND, NOR, XOR, XNOR).

A simple implication is a local evaluation of a given gate. Additionally, it is common to distinguish between *forward* and *backward* implication, depending on whether the new value assignment is implied at the inputs or the output of the considered gate. Figure 2.4 shows examples of simple forward and backward implications.

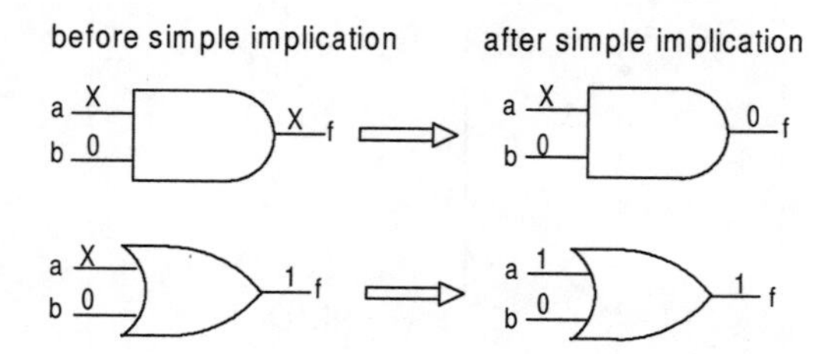


**Figure 2.4:** Simple forward- (upper part) and simple backward (lower part) implications

> **Definition 2.3:** A logic value assignment is obtained by *direct implication*, if it is determined by a sequence of simple implications.

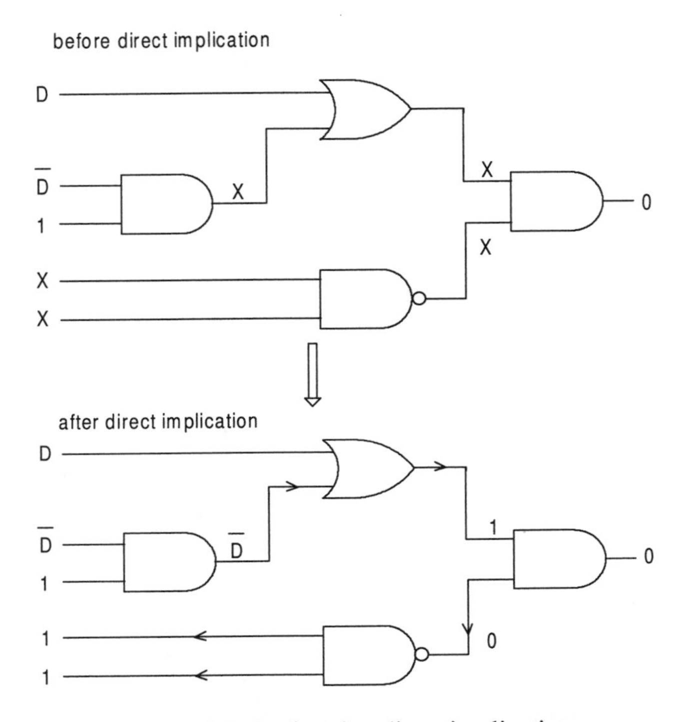


**Figure 2.5:** Performing direct implications

While simple implications are restricted to locally evaluating single gates, direct implications can make value assignments across several gates along the paths in a circuit. Also, it is sensible to distinguish between forward and backward implications depending on whether the signal values are propagated towards the primary outputs or primary inputs, respectively. Figure 2.5 shows an example for the performance of direct implications.

**Definition 2.4:** If a value assignment at a given signal is uniquely determined by previous value assignments in the circuit and cannot be implied directly, then it is determined by *indirect* implication.

Figure 2.6 shows an example of an indirect implication proposed by Schulz et al. [ScTr87] which will be studied again in later sections.

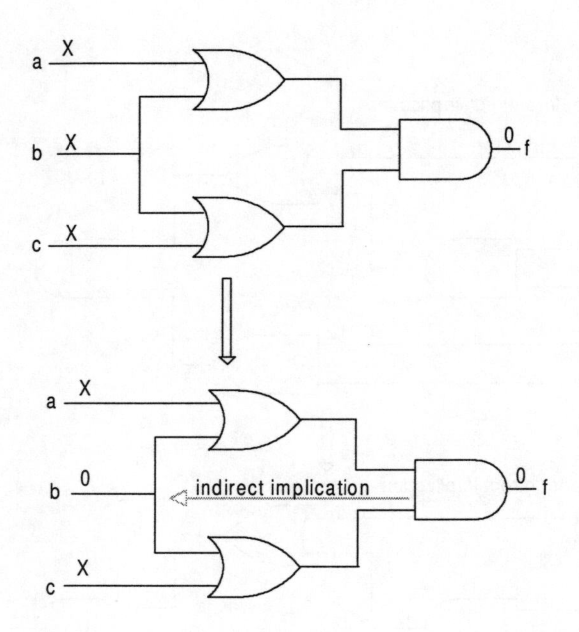


**Figure 2.6:** Example of indirect implication [ScTr87]

As will be explained in more detail, the implication procedure plays a central role in all test generators. The large majority of test generators for combinational circuits like the D-algorithm [Roth66], PODEM [Goel81], FAN [FuSh83], TOPS [KiMe87] exclusively perform simple and direct implications. This can be done easily by evaluating the truth tables of every gate that has an event and by propagating the signal values in an event-driven way. Unfortunately, no indirect implications can be performed using these concepts. The reader may verify this by means of Figure 2.6. For the example in Figure 2.6, it is possible to derive a uniquely determined value although no simple implication can be performed. In Chapter 3 a method will be proposed that allows us to perform *all* indirect implications in a circuit.

## 2.4 General Procedure

In spite of the large variety of combinational test generation tools there exist a lot of concepts common to many of them. This section explains and illustrates the general combinational ATPG-procedure along with some basic heuristics employed by many tools. Most heuristics are derived from *structural* properties of the circuit under test. It is the great merit of Roth's D-calculus that it allows us to completely describe the faulty behavior of a circuit on its structural gate-level description. Therefore, all operations needed for test generation can be performed on a structural circuit description paving the way for efficient heuristics. In this chapter, we exclusively look at such structural techniques being derived from Roth's D-algorithm. They have widely replaced earlier algebraic techniques based on *Boolean difference*. For more details on Boolean difference, see e.g. [AbBr90]. With some notable exceptions, like the Boolean satisfiability method [Larr89], the transitive closure method [ChAg93] or BDD-based approaches, e.g. [StBh91], almost all modern tools follow the basic procedure to be presented.

For illustration of the general ATPG-procedure consider the circuit of Figure 2.7.

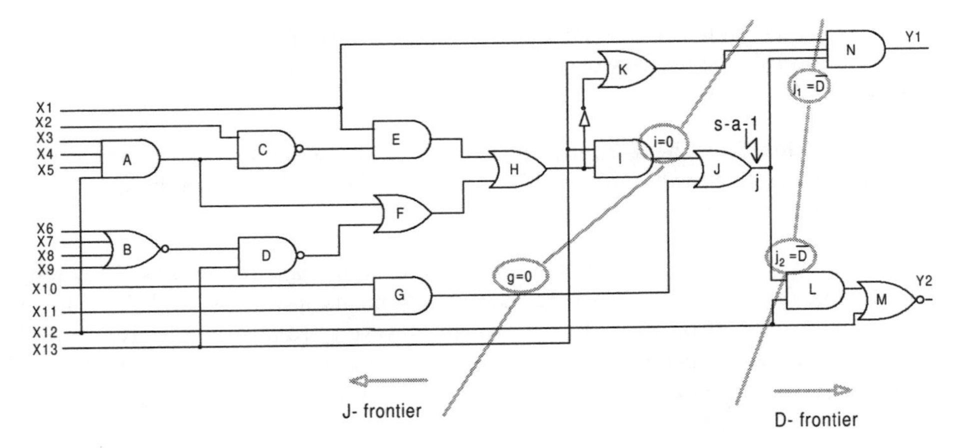


**Figure 2.7:** Fault injection

Consider the single stuck-at-1 (s-a-1) fault at signal $j$ in Figure 2.7. It is the task of the test generator to find a set of assignments for the input signals of the circuit such that this fault causes a faulty response at least on one output of the circuit. An input vector can only be a test for this fault if it generates a faulty logic value or *fault signal* at signal $j$. In our example, signal $j$ has to be *controlled* from the inputs in such a way that it assumes a logic 0 in the fault-free case. If signal $j$ is 0 in the fault-free case and 1 in the faulty case, then it assumes the logic value $\bar{D}$ in Roth's notation. Note that $j = \bar{D}$ represents a *necessary value assignment* as there exists no set of value assignments in the circuit without $j = \bar{D}$ which still detects the fault at least on one primary output. Therefore, all test generators start with this value assignment, also referred to as *fault injection*.

After fault injection all common test generators perform logic implications, i.e., they make value assignments that can be derived from fault injection. In Section 2.3 different types of implications were defined and it depends on the specific test generator which types of implications are actually performed. In any event, the implication procedure produces new value assignments. In the example of Figure 2.7 all *direct* implications (forward and backward) have been performed. This results in a new situation of value assignments being the starting point for all subsequent test generation steps. Among the signal assignments generated during the ATPG-process, there are two sets of signals that are of particular interest:

> **Definition 2.5:** The *D-frontier* (*D-drive*) *F* consists of all fault signals, i.e., signals being assigned either D or $\overline{\text{D}}$, which are input signals of logic gates whose output signal is unspecified.

The *D-frontier* indicates how far the faulty signals have propagated from the fault location towards the primary outputs. In Figure 2.7 fault injection has produced the D-frontier $F = \{j_1, j_2\}$.

> **Definition 2.6:** The *J-frontier* *U* consists of all output signals of gates being assigned a fixed value, where this logic value cannot be obtained by *simple implication* from the gate inputs.

More commonly, the signals of the J-frontier are called *unjustified lines*. They represent signals in the circuit, where value assignments have produced fixed logic values at the ouputs of internal gates which, however, are not completely *justified* by the value assignments at the gate inputs. In Figure 2.7 two unjustified lines $U = \{i, g\}$ have been created. For a valid test, they have to be justified in subsequent steps of test generation.

In short, the task of a test generator is formulated as follows: find a set of binary value assignments at the primary inputs of the circuit, such that

- the *D-frontier* reaches the primary outputs, i.e., at least one primary output assumes a faulty value,
- the *J-frontier* reaches the primary inputs, i.e., there exist no unjustified lines in the interior of the circuit.

To accomplish these tasks most algorithms of deterministic test generation proceed step by step, assigning appropriate logic values at well-selected signals in the circuit such that the D-frontier is moved towards the primary outputs and the J-frontier is moved towards the primary inputs. This is shown schematically in Figure 2.7.

Along the process of test generation we distinguish two types of value assignments. Firstly, we consider *necessary assignments*. Necessary assignments are uniquely determined by the current situation of value assignments in the circuit and any test vector that can be generated starting from the current situation *must* contain this value assignment. In Figure 2.7 fault injection has produced the necessary value assignments $i = 0$, $g = 0$, $j_1 = \overline{D}$ and $j_2 = \overline{D}$. Besides necessary assignments every test generator makes *optional assignments*. Optional assignments are assignments that *can* be made in order to reach the goal of generating a test vector. The existence of optional assignments results from the fact that there usually exist more than just one test vector for a given fault. By making optional assignments we continue to restrict the number of possible test vectors until we finally end up with exactly one. The choice of optional value assignments is subject to *heuristics*.

Test generation can be understood as a sequence of making optional and necessary assignments. How necessary assignments are identified and what heuristics are used to select optional assignments depends highly on the individual test generation method and determines the power of the different ATPG-tools. Nevertheless, the general procedure is nearly the same for most conventional methods. The general program flow shown in Figure 2.8 is typical of a large majority of presently popular tools.

The steps of test generation summarized in Figure 2.8 shall be briefly explained:

*A) Fault injection (fault set-up)*

The fault signal at the fault location immediately results from the fault value (stuck-at-1 or stuck-at-0) and the definition of the underlying logic alphabet. In Figure 2.7 this was illustrated for 5-valued logic.

*B) Unique sensitization*

After fault set-up most tools conduct a topological analysis examining the paths along which the fault signal can possibly propagate. Certain topological concepts are used to derive value assignments being necessary for the propagation of the fault signal. Such *unique sensitization* techniques are subject of Section 2.5.

*C) Implications*

For the previously made value assignments implications are performed. At this point, tools may differ in their ability to perform different types of implications according to Section 2.3. For example in PODEM [Goel81] only direct forward implications are performed while FAN [FuSh83] additionally takes direct backward implications into account. It is the particular contribution of SOC-

RATES [ScTr87] that for the first time indirect implications were performed in special cases. (They were called *global implications* in [ScTr87].) In [KuPr92] a test generator is presented that can perform *all* indirect implications. This complete implication technique will be subject of Chapter 3.

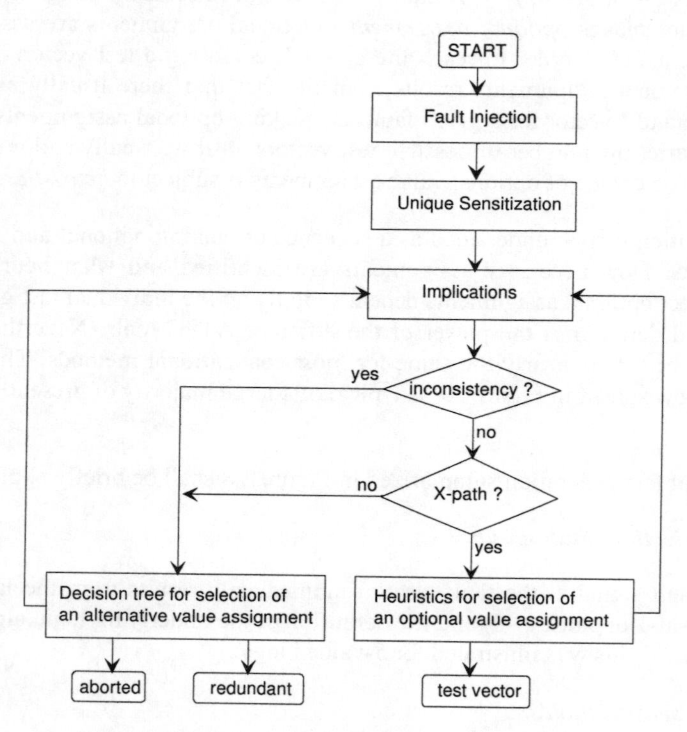


**Figure 2.8:** General procedure of test generation

*D) Checking*

Another situation of value assignments forms the basis for the next steps in test generation. First however, it has to be examined whether the current situation still allows for the generation of a test vector. This is generally not an easy task, however, there are two conditions that must be fulfilled at any moment during the test generation process and these can easily be checked:

- logic consistency
- existence of an *X-path*

First, the logic value assignments must be consistent, i.e., there must be no logic contradictions between signal values in the circuit. Usually, logic consistency is verified during the implication process. Second, there has to exist at least one path in the circuit along which the fault signal can propagate to at least one of the primary outputs. This condition is fulfilled if there is a path from at least one signal of the D-frontier to one of the primary outputs along which all signals are unspecified, i.e., they have the logic value X. We say that the current situation passes the *X-path-check* [Goel81]. Of course, the X-path-check only has to be conducted as long as the fault signal has not yet reached one of the primary outputs. Generally, we speak of a *conflict* if either the X-path-check has failed or logic inconsistencies have occurred.

*E) Optional value assignments*

If no conflict has occurred and no test vector has been obtained yet, heuristic methods are employed to suggest optional value assignments. There is a large variety of techniques that have been reported in the literature for this purpose. Roughly, common heuristics for the selection of optional value assignments can be divided into two categories:

- testability measures
- backtrace-procedures

Generally, *testability measures* have the task to estimate the probability that a given fault is tested by a *random* input vector. Testability measures are usually composed out of *controllability measures* and *observability measures*. Controllability measures estimate the probability for a given (fault-free) signal to assume the logic value 0 or 1. Observability measures predict the probability of a given fault signal to propagate to one of the primary outputs. In test generation, observability measures are therefore used to select a "good" signal from the D-frontier and a "good" path along which this signal can be propagated. Controllability measures are used to obtain smart strategies for the justification of the signals in the J-frontier. For example the following testability measures are commonly used: COP [Brgl83], SCOAP [Gold80], STAFAN [JaAg85], LEVEL [Lioy87].

However, experience shows that testability measures have only limited impact on the performance of a test generator. The most simple testability measures, (e.g., controllability and observability of a signal is simply derived by its topological distance from primary inputs and primary outputs, respectively) have shown to be sufficient for the purpose of test generation, in the sense that more sophisticated testability measures do not lead to substantially improved results.

The performance of test generators, however, can be greatly influenced by the use of *backtracing* (not to be confused with *backtracking*) procedures. Starting from the signals in the J-frontier and D-frontier, backtracing procedures trace through the different paths of the circuit towards primary inputs. This process is guided by testability measures and signals are identified where certain value assignments seem wise. Generally, those signals are identified for which the processed information indicates a risk of a conflict [FuSh83]. If no such signals exist, information from backtracing is used to select value assignments that are particularly promising in order to quickly drive the test generation process towards certain *objectives* [FuSh83]. The objectives result from the general strategy to move the D-frontier towards primary outputs and the J-frontier towards primary inputs. The reader may study in more detail the multiple backtrace procedure of [FuSh83] which essentially follows the above described philosophy and, to this date, represents an integral part of almost all modern ATPG-tools.

Importantly, only *one* and not several *optional* value assignments are made at a time. After each optional value assignment, implications have to be performed again. The only exception known to the authors is the test generator in [GiBu91] where under certain conditions several assignments can be made at once. This can yield some speed-up for easy to test faults but does not have any impact on the performance of test generation for difficult to test faults and redundancy identification.

*F) Decision tree (backtracking)*

In Section 2.2 test generation has already been presented as a search in the finite Boolean space. The search consists of making decisions that are added to the decision tree. In test generation, *optional* value assignments constitute decisions. Each decision is associated with a node in the decision tree. Importantly, *necessary* assignments are not decisions and they are therefore not represented in the decision tree. If a conflict occurs during test generation, previous decisions have to be reversed. As explained in Section 2.2 *backtracking* can be done systematically by means of the decision tree. The decision tree guarantees that the search is *complete.* For every fault, if given enough time, either a test vector is generated or it is proved *untestable* (i.e., *redundant*). Unfortunately, this search can sometimes consume unacceptable amounts of computation time and all test generators therefore use an *aborting criterion.* It is common to abort the search after a certain number of backtracks. This number is typically chosen between ten and a few hundreds. If the search for a given fault could not be completed, we speak of an *aborted fault*. The performance of a test generator is therefore evaluated not only in terms of the required CPU-time but also according to the number of faults aborted in a given circuit.

## 2.5 Topological Analysis for Unique Sensitization

As explained in Section 2.4, there are two issues during test generation that require attention. First, it has to be ensured that the logic situation of value assignments is always consistent. Logic inconsistencies have to be identified as early as possible. This is the task of implications as introduced in Section 2.3 and will be further discussed in Section 2.7 and Chapter 3. The second goal of the algorithm is to make sure that at least one path exists in the circuit along which the fault can propagate to a primary output. It is therefore crucial to calculate necessary conditions for fault propagation. Often it is possible to derive certain necessary conditions simply from the *topology* of the circuit.

All topological examinations for *unique sensitization* (to be explained) are based on a representation of the circuit $C$ as a *directed acyclic graph* (DAG), $S = (V, E)$. The vertex set $V$ contains all primary inputs of the circuit as well as all output signals of gates with at least two inputs. (Inverters are usually not represented in the graph.) The set of edges $E$ contains all directed edges of the graph. Thus, $(a, b) \in E$, if signal $a$ is input of a gate in the circuit with output signal $b$. Figure 2.9, as an example, shows the directed acyclic graph of the circuit introduced in Figure 2.7. The vertex set $V$ for the circuit of Figure 2.7 is given by $V = \{x_1, x_2,\dots x_{13}, a, b,\dots n\}$ and the set of edges is $E = \{(x_1, n), (x_1, e), (e, h), (x_2, c),\dots \}$.

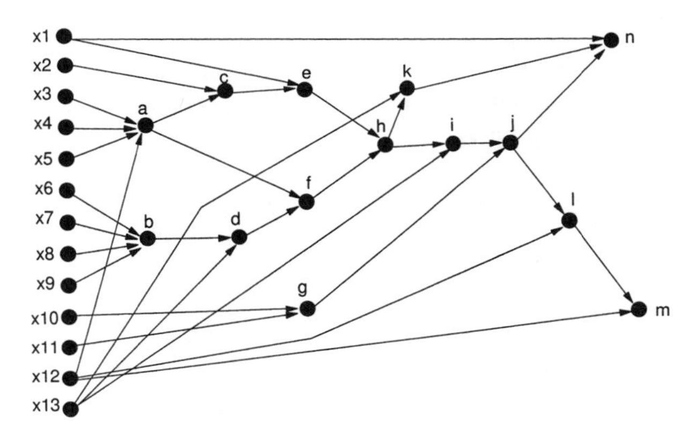


**Figure 2.9:** Representation of the circuit in Figure 2.7 as *directed acyclic graph*

Often, it is important to determine whether a signal is structurally *reachable by the fault signal*.

> **Definition 2.7**: Consider some fault at an arbitrary signal $f$ in the circuit. An arbitrary signal $r$ in the circuit is said to be *reachable by the fault signal* if there exists a directed path from $f$ to $r$. If no such path exists, signal $r$ is *unreachable by the fault signal.*

The information whether or not a given signal can be reached by the fault signal plays an important role when choosing a value assignment. A signal $r$ not reachable by the fault can only assume the fixed logic values 0 and 1. Otherwise, if $r$ is reachable by the fault effect, the value set $\{0, 1, D, \bar{D}\}$ has to be considered.

In order to better describe the process of propagating a fault signal it is useful to distinguish between the *controlling value* and the *non-controlling value* of a given gate type:

> **Definition 2.8:** Let $g$ be a gate with the unspecified output $y$ and $n$ unspecified inputs $x_1, x_2,... x_n$, $n \geq 2$. If the value assignment $x_i = V$, with $V \in \{0, 1\}$ and $i \in \{1, 2,... n\}$, uniquely implies $y = W$, $W \in \{0, 1\}$, then $V$ is called the *controlling value* of gate $g$. Otherwise, it is called the *non-controlling value* for $g$.

Applying this definition to specific gate types we realize that 0 is the controlling value for AND and NAND, and 1 is the controlling value for NOR and OR. In the case of XOR and XNOR, both 0 and 1 are non-controlling values.

> **Example 2.1:** Figure 2.10 shows an AND-gate and a NOR-gate, each having two inputs $x_1$ and $x_2$ and an output $y$. Assume in both cases that signal $x_1$ cannot be reached by the fault signal and that the fault signal D at input $x_2$ shall be propagated to output $y$. This is only possible if the considered gate is *sensitive* with respect to the input $x_2$, i.e., a signal change at $x_2$ also causes a signal change at $y$. Studying the truth tables of the D-calculus in Table 2.1 it is easy to see that a gate can only be sensitive, if all inputs without the fault signal assume the non-controlling value.

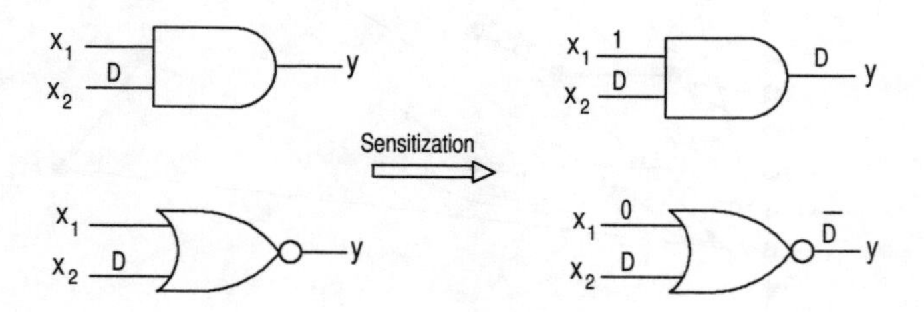


**Figure 2.10:** Sensitization of gates

**Definition 2.9:** A gate $g$ is called *uniquely sensitized* [FuSh83] with respect to an input $x$ if all values that are necessary to propagate a fault signal from $x$ to the output of $g$ are assigned at the inputs of $g$.

Note that a gate can only be uniquely sensitized if all inputs that are unreachable by the fault effect are assigned a non-controlling value. These necessary assignments can be identified very easily because only a structural analysis is required to check what signals are reachable. However, as will be illustrated in Section 2.6, this simple analysis does not identify *all* necessary assignments for fault propagation. A complete method for sensitization will be presented in Section 3.2.2.

It remains to be examined what gates have to be sensitized in order to permit fault detection. As can be noted, all gates *have to be* sensitized through which the fault signal *must* propagate in order to reach one of the primary outputs. This leads to an important application of the notion of *dominators* in directed acyclic graphs in the field of test generation [KiMe87].

**Definition 2.10:** A node $n$ *dominates* node $m$ in a directed acyclic graph $S$, if and only if every directed path from node $m$ to any terminal node of $S$ contains node $n$. Node $n$ is then called *dominator* of $m$ in graph $S$.

In test generation, it is very helpful to identify the dominators of a given fault signal. They represent the bottlenecks for fault propagation. Clearly, if node $g$ in the DAG of $C$ is a dominator for some node $f$, then a fault signal starting at fault location $f$ in circuit $C$ *must* pass through gate $g$ in order to reach a primary output of $C$. Using this concept the algorithms FAN [FuSh83] and TOPS [KiMe87] perform *unique sensitization*. By unique sensitization we understand the process of making value assignments that are necessary to propagate the fault signal to at least one output of the circuit. The procedure in [FuSh83] and [KiMe87] can be summarized as follows:

*Unique sensitization* [FuSh83], [KiMe87]: Consider a circuit $C$ with a fault location $f$. For all gates in the circuit whose output signals are dominators for signal $f$, assign the non-controlling value to all inputs that are unreachable by the fault effect.

**Example 2.2:** As an example, consider again the graph in Figure 2.9. For a target fault at signal $b$ we obtain dominators $d$, $f$ and $h$. For the circuit in Figure 2.7 this results in the necessary assignments for unique sensitization $x_{13} = 1$, $a = 0$ and $e = 0$. Now consider the case where a target fault is located at gate $a$. The only dominator is $h$ suggesting that the corresponding gate has to be sensitized. However, in this case, this does not result in any unique value assignment since signals $e$ and $f$ are both reachable by the fault signal.

Further topological concepts have been derived form the notion of dominators. In [MaGr90] it is proposed to partition the circuit into *cones*. This means that all considerations are restricted to one primary output at a time and each primary output is separately processed. This results in a larger number of *local* dominators in each cone leading to a considerable acceleration of the test generation process.

## 2.6 The Problem of ATPG: An Example

In order to gain better insight for how we run into problems during test generation reconsider the circuit example of Section 2.4 which is repeated for convenience in Figure 2.11. Following the general test generation scheme of Section 2.4 we attempt to generate a test vector for the shown fault.

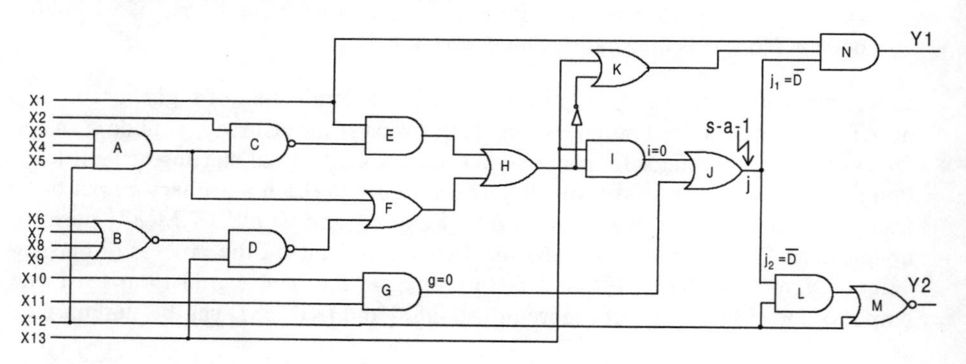


**Figure 2.11:** A test generation example

It depends on the details of the employed heuristics how exactly the test generation algorithm proceeds. In this example, assume that the implication procedure is able to perform direct implications only. Like in PODEM [Goel81] optional value assignments are made exclusively at the primary inputs. In this example, value assignments are chosen as they seem "reasonable" when "looking" at the circuit. In doing so, to keep things simple, we imitate common heuristics in a simple way.

Table 2.3 depicts the single steps of test generation, i.e., it lists the decisions as a typical test generator would make them to generate a test for the shown fault in the circuit of Figure 2.11. First fault injection is performed at the location of signal *j* as shown in the figure. There are no dominators for signal *j* and hence no additional assignments result from unique sensitization. Next, as shown in the figure implications are performed resulting in the shown D-frontier and the unjustified lines $i = 0$ and $g = 0$. Since there is no logic inconsistency and the situation passes the X-path-check, we proceed with an optional value assignment.

Some heuristics may suggest that the fault signal should be propagated through gate $N$. This is reasonable because primary output $y_1$ is closer to the fault location than $y_2$ and typically this would also be identified by some testability measure. Hence, assume that gate $N$ has to be sensitized. Therefore, assignment $x_1 = 1$ seems promising because it will work towards sensitizing $N$ and we choose this assignment as our first decision. Sensitization of $N$ however is only one worthwhile objective. Also, we have to make sure that the unjustified lines become justified. Looking at $i = 0$ we realize that, e.g., $h$ should eventually become 0 for the purpose of justifying line $i$, and by *backtracing* we find that $x_2 = 1$ contributes to this goal. It therefore becomes our second decision. For similar reasons we will take the other decisions as shown in Table 2.3. As explained in Section 2.4, after each optional value assignment, implications are performed and the resulting situation is checked for conflicts. The implications resulting after every optional value assignment are also shown in Table 2.3. After the eleventh decision, i.e., after value assignment $x_{12} = 1$, a conflict has occurred. The X-path-check has failed since there is no primary output that can be reached by the fault signal. If a test vector exists at all, then it cannot be based on the current set of value assignments and backtracks have to be performed. By means of a decision tree all alternative value assignments are explored systematically step by step.

A closer study of this example circuit reveals however that no test exists for the considered fault. The fault is redundant. Therefore, the whole decision tree must be exhausted until that is proved. For large trees this enumeration process can consume unacceptable amount of computation time. Even in this small example, 11 nodes have been added to the decision tree and in the *worst case*, $2^{11} = 2048$ value assignments have to be performed to exhaust all possibilities. Since the decision tree is a method for *implicit* enumeration this number can be smaller if *non-solution areas* of the decision tree are recognized at an early stage. However, the example shows how a test generation algorithm can run into severe problems even for small circuits.

This illustrates how the practical performance of a test generator sometimes can indeed be close to the exponential worst-case behavior as it results from the NP-completeness of the fault detection problem. To avoid such worst case behavior as often as possible, the central issue in test generation can be summarized as follows:

> *What techniques allow to efficiently detect the non-solution areas of the search space in practical examples?*

In the above example, it would be ideal to have a technique which determines that the situation is conflicting and hence no test exists, already *before any optional* decision (they are all wrong!) has been taken. We study Figure 2.11 again to understand why the considered fault is redundant.

| Decision | Optional value assignment | Implications |
|---|---|---|
| 1 | $x_1 = 1$ | |
| 2 | $x_2 = 1$ | |
| 3 | $x_3 = 1$ | |
| 4 | $x_4 = 1$ | |
| 5 | $x_5 = 1$ | |
| 6 | $x_6 = 0$ | |
| 7 | $x_7 = 0$ | |
| 8 | $x_8 = 0$ | |
| 9 | $x_9 = 0$ | $b = 1$ |
| 10 | $x_{10} = 0$ | $g = 0$ (justified) |
| 11 | $x_{12} = 1$ | $a = 1$; $c = 0$; $e = 0$<br>$f = 1$; $h = 1$;<br>$x_{13} = 0$ (since $i = 0$)<br>$k = 0$ ; $y_1 = 0$;<br>$l = \overline{D}$ ; $y_2 = 0$ |

**Table 2.3:** Steps of test generation in circuit example of Figure 2.11

Observation 1:

Assume that the fault signal shall be propagated to primary output $y_2$. This requires that both gates $L$ and $M$ are sensitized. To sensitize $L$ the assignment $x_{12} = 1$ is necessary. However, this assignment desensitizes $M$ and the primary output $y_2$ assumes the logic value 0. Consequently, the fault signal can *never* be detected at output $y_2$. Hence, the fault can only be detected at primary output $y_1$ and gate $N$ has to be sensitized. This leads to the unique sensitization assignments $x_1 = 1$ and $k = 1$.

It is important to note that these unique sensitization assignments cannot be identified by a *topological* analysis alone. Additionally, in the illustrated situation, a *logical* analysis is unavoidable.

Observation 2:

Reconsider the value assignments $x_1 = 1$ and $k = 1$ found in observation 1. Direct implications cannot derive any additional assignments from this. A closer study of the example shows however that because of $x_1 = 1$ the assignment $h = 1$ is necessary. The reader may realize this by assuming the assignment $h = 0$ and pursuing its direct implications. Since $x_1 = 1$ we obtain a logic inconsistency for $h = 0$ at signal $a$ so that $h$ must be 1. At this point we would need a technique that performs the *indirect* implication ($x_1 = 1 \Rightarrow h = 1$). If we now perform the direct implications for $h = 1$ we obtain the necessary assignments $x_{13} = 0$ and $k = 0$ (because of $i = 0$). This however is logically inconsistent with the unique sensitization $k = 1$ resulting from observation 1. It can be concluded that even without any optional value assignments there is a conflict and the fault is therefore not testable.

In order to pursue the reasoning of observation 2 it is required to also perform *indirect* implications. The conventional direct implication procedures, in the above example, are not able to yield the conflict. Again we note that only a more detailed *logical* analysis of the circuit can solve the problem.

Realize that the ability of an algorithm to identify non-solution areas of the search space solely depends on its methods to identify necessary value assignments. In correspondence to the above two observations, in a FAN-based test generation environment two types of necessary assignments are distinguished: the first type consists of the value assignments resulting from the necessity to propagate the fault signal to at least one primary output. We also refer to this type as unique sensitization. It has been illustrated in observation 1. Observation 2 illustrates the second type. This type corresponds to the cases where a value assignment is necessary for the logic consistency of the given situation of value assignments. This type of necessary assignments has to be found by the implication procedures.

Motivated by these observations in [KuPr94] a method called *recursive learning* has been proposed to calculate *all* necessary assignments of both types. This technique is subject of Chapter 3.

## 2.7 Static and Dynamic Learning

In [ScTr87], the so called *learning strategies* have been proposed for identifying necessary assignments during test generation. Learning techniques conduct a *logical* analysis of the circuit. Schulz et al. [ScTr88] proposed *static* and *dynamic learning* which is able to identify certain necessary assignments not identified by previous implication methods.

In order to identify as many necessary assignments as possible, each given situation of value assignments has to be considered individually. If *learning* operations are conducted after every step of test generation, i.e., after each optional value assignment, this is called *dynamic learning* [ScAu89]. Often however, test generation for different faults leads to the same situation in the circuit and the same implications can be performed. Therefore it is of advantage to conduct certain learning operations already *before* the actual test generation process is invoked. All information learned is then available for all subsequent steps of test generation. Learning being conducted in such a *pre-processing phase* is referred to as *static learning* [ScTr87].

In static learning, the immediate goal is not to identify necessary assignments since there is neither a given set of value assignments nor a specific fault considered. However, during pre-processing, certain logic relations between signals in the circuit can be examined. It is the goal of static learning to identify as many generally valid indirect implications in the circuit which can be stored so that they are available when needed. This storing of implications has probably been the original motivation for the name "Learning".

In later techniques like dynamic learning [ScAu89] and recursive learning [KuPr92], storing of implications is only one aspect and may only be used as an option. Generally, in the field of computer-aided circuit design the purpose of these "learning" techniques is the automatic acquisition of implications in a combinational network .

> *Learning* refers to the process of *monitoring an enumeration of value assignments* with the goal of *extracting implications or necessary assignments* by certain *inference rules*.

The main difference between the techniques by Schulz and Auth [ScAu89] and recursive learning [KuPr94] lies in the different nature of the enumeration schemes being monitored by the learning method. The method of Schulz et al. is based on conventional *variable enumeration* as is standard in many simulation techniques. Recursive learning is based on a fundamentally different enumeration that has been called *AND/OR enumeration*. This will be further developed in Chapter 3 and 4.

The following learning method based on conventional variable enumeration has been proposed by Schulz et al. [ScTr87] and is shown in Table 2.4. The procedure in Table 2.4 can be applied statically and dynamically. It assigns the logic values 0 and 1 after each other at every signal in the circuit and examines their logical consequences. That is, it monitors the enumeration of a single variable at each node in the circuit. For this purpose procedure *implications()* performs direct implications along with known indirect implications that may have been stored before. Procedure *analyze_result()* examines whether worthwhile information can be derived from the performed steps by certain rules. Information about necessary assignments is considered worthwhile learning if these assignments cannot be derived by direct implications. This shall be explained in more detail.

```
/*  this routine operates on a global data structure repre-
    senting the gate netlist of the circuit */

Learning()
{
    for (every signal f in the circuit)
    {
        value assignment in the circuit, f := 0;
        implications();
        analyze_result(f);

        value assignment in the circuit, f := 1;
        implications();
        analyze_result(f);
    }
}
```

**Table 2.4:** Learning procedure of Schulz, Trischler and Sarfert [ScTr87]

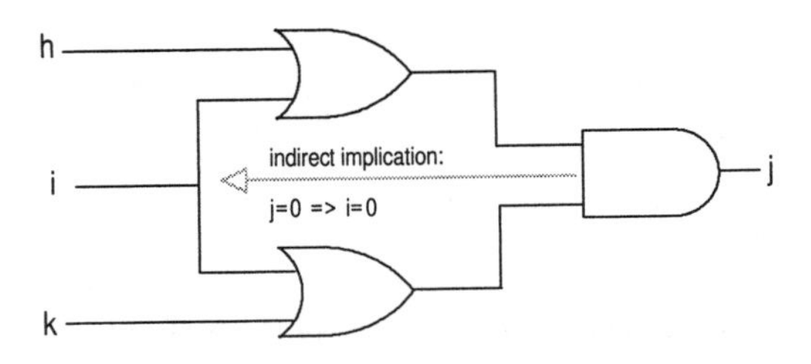


**Figure 2.12:** Static learning of indirect implication

Figure 2.12 shows a circuit already discussed in Section 2.3 illustrating that direct implications alone are not sufficient to identify all necessary value assignments. A closer look reveals that the indirect implication $j = 0 \Rightarrow i = 0$ is true. Assume that the circuit is part of a larger circuit and for didactic reasons we first consider the case of dynamic learning. Assume that during test generation value assignments have been made somewhere in the circuit and by implications signal $j$ in Figure 2.12 has taken the value 0. Clearly, in the shown circuitry the current situation of value assignments consists of a value assignment $j = 0$ for which direct implications do not produce any necessary assignments. Now we perform (dynamic) learning using the procedure of Table 2.4. At some point, signal $i$ is picked and the value 1 is assigned. Implications are performed (directly) and we obtain $j = 1$, and hence a conflict is produced. This means that only the value assignment $i = 0$ can be consistent with the present situation. The case of a logic inconsistency is covered in *analyze_result()* by a *learning rule* [ScTr87] and the necessary assignment $i = 0$ is learned. For dynamic learning, Schulz et al. have proposed four learning rules contained in *analyze_result()*. The interested reader may refer to [ScTr87], [ScAu89].

```
/*  this routine operates on a global data structure representing the gate netlist of the
    circuit */

analyze_result(i)
{
    /* learning rule 1: for static and dynamic learning */
    Vi := value of signal i in the circuit;
    for (every signal j ≠ i with logic values 0 or 1 where j is input or output of a gate
    g with at least two inputs being unspecified before the value assignment i := Vi )
    {
        Case 1: (j is output and x is input of g) and
                (the simple forward implication x = Vx ⇒ j = Vj has been performed)
                and (Vx is non-controlling value of g) then:

                    Store the implication: j = V̄j ⇒ i = V̄i

        Case 2: (j is input and y is output of g) and
                (the simple backward implication y=Vy ⇒ j=Vj has been performed)
                and (Vj is controlling value of g) then:

                    Store the implication: j = V̄j ⇒ i = V̄i
    }   /* end of learning rule 1 */

    /* learning rule 2-4: for dynamic learning only, see [ScAu89] */
}
```

**Table 2.5:** Learning criterion

Now consider static learning for the above example. During the pre-processing phase all signal values are still unspecified and the objective is not to determine necessary assignments but to identify and store indirect implications. Again, we use the learning procedure of Table 2.4 for signal $i$. For $i = 1$ we obtain $j = 1$ by direct implication. For this implication, it is possible to apply the following basic logic law. For arbitrary signals $i, j$ in the circuit and arbitrary logic values $V_i$, $V_j \in \{0, 1\}$, it holds:

*Law of contraposition:*

If $i = V_i \Rightarrow j = V_j$ is true, then $j = \overline{V_j} \Rightarrow i = \overline{V_i}$ is also true.

In the above example, by means of contraposition, the indirect implication $j = 0 \Rightarrow i = 0$ is easily derived from the direct implication $i = 1 \Rightarrow j = 1$. It is reasonable to store the indirect implication at signal $j$ so that it can be used whenever $j$ assumes the value 0. Therefore the law of contraposition should be included in procedure *analyze_result()* as one of the learning rules. This creates a difficulty: Most implications being derived from direct implications by contraposition are also direct and are therefore not worthwhile learning. A criterion is needed to decide in a quick and reliable way whether the derived implication is direct or indirect. For this purpose, a *learning criterion* has been proposed in [ScTr87]. Table 2.5 shows an extension to this learning criterion proposed in [KuPr93].

The reader may verify that *Case 1* of the learning criterion is fulfilled for the example in Figure 2.12. The assignment $i = 1$ directly implies 1 at both inputs of the gate with output $j$. Since 1 is the non-controlling value for AND gates and because $j = 1$ is implied by simple forward implication all conditions for *Case 1* are fulfilled and the corresponding indirect implication is stored. *Case 1* of the above criterion corresponds to the learning criterion in [ScTr87]. In the original criterion of Schulz, cases of indirect forward implications can be missed. They are taken into account in the above criterion by *Case 2*.

> **Example 2.3:** For illustration consider Figure 2.13. Suppose *learning()* is performed for the circuit in Figure 2.13. For the assignment $i = 0$ direct implications yield $d = 0$ and $f = 0$. Furthermore, it can be directly implied that $b = 0$ and $e = 1$. Since $f = 0$ we also imply $j = 0$. Note that the simple implication yielding $j = 0$ fulfills all conditions for *Case 2* and the indirect implication $j = 1 \Rightarrow i = 1$ is stored.

Importantly, in static learning, the implication $j = \overline{V_j} \Rightarrow i = \overline{V_i}$ is already identified *before* signal $j$ has actually assumed value $\overline{V_j}$. The indirect implication is stored during pre-processing at signal $j$ and can always be used, when the test generator for some arbitrary fault assigns $j = \overline{V_j}$ and *if neither signal i nor signal j can be reached by the fault signal.* This restriction is needed because the law of contraposition can only be applied in the way shown if the given signals only assume values 0 or 1. If a signal can be faulty more logic values have to be considered and $\overline{V}$ would not represent a unique value. Unfortunately, this restriction considerably limits the use of the stored implications.

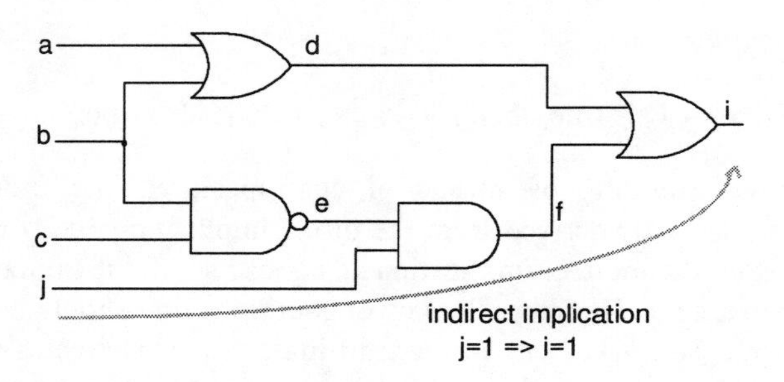


**Figure 2.13:** Indirect forward implication

In dynamic learning, a lot more information can be obtained about the circuit. This is due to the fact that many indirect implications needed in specific situations to recognize necessary assignments are actually only valid in these specific situations. Therefore, to learn as many necessary assignments as possible, the learning process has to be repeated for every situation of value assignments, i.e., after every optional decision during the test generation process. For dynamic learning, unlike in static learning, there are specified signals in the circuit so that more cases of "interesting" learning situations can occur. Therefore, besides the learning rule shown in Table 2.5, procedure *analyze_result()* has been extended by three additional learning rules [ScAu89].

While dynamic learning can usually extract a lot more information about the circuit than static learning, it on the other hand requires substantially more CPU-time because all learning operations have to be repeated after every decision. In [KuPr93] a dynamic learning technique was proposed being able to update the information after every step incrementally so that significant speed-up is gained. However, even then the required CPU-time is large and the gained information does not always justify the extra effort. Therefore, in most state-of-the-art test generators *dynamic* learning is not employed.

Instead of investing great effort into calculating necessary assignments it is also possible to refine the backtracking process that is unavoidable if necessary assignments are missed. This has been studied by Giraldi, Bushnell and Chen [GiBu91], [ChBu96], who proposed an efficient procedure to identify equivalent search states during test generation. Storing of search states requires additional memory but it can significantly reduce the computation times for backtracking if equivalent search states are identified. If search states are stored each portion of the search space needs to be visited only once. In [FuFu92] this approach has been extended to also take into account dominance relations between search states.

The techniques described here assume that the circuit is represented by a combinational network as defined in Section 1.3.3. It should be noted however that powerful test generation techniques also exist for other circuit representations. Larrabee proposed a representation of the circuit by a set of binary and ternary clauses and presented an efficient test generation method based on Boolean difference [Larr89]. Chakradhar, Agrawal and Rothweiler [ChAg91], [ChAg93], developed a powerful implication procedure for this representation that also takes into account ternary relations. In [ChAg91], [ChAg93] the set of clauses is obtained from a neural network model [ChBu91] and necessary assignments for test generation are calculated based on the transitive closure of an implication graph. An efficient implication graph for clause-based circuit representations has also been proposed in [HeWi95].

# Chapter 3

# RECURSIVE LEARNING

This chapter describes an algorithm for identifying *all* necessary assignments for single stuck-at fault detection. This is of central interest not only in test generation but also in logic synthesis and verification as will be developed in Chapters 5 and 6.

Since the original motivation for this work was to improve the deterministic portion of test generation algorithms, in this chapter, the basic notions of recursive learning are introduced with test generation as background. As explained in Section 2.6, complete knowledge about necessary assignments at every search state provides exact identification of the non-solution areas. If all necessary assignments are known in a given search state no wrong decision can be made, and backtracking can be completely avoided. Hence, the test generation problem is solved if all necessary assignments can be identified.

This chapter is dedicated to giving a technical description of the algorithm and to illustrate the procedure by various examples. A more global view of this technique is developed in Chapter 4 where recursive learning is generalized by introducing *AND/OR reasoning graphs*.

The chapter is organized as follows: for a quick orientation of the reader, a simplified recursive learning technique is introduced and discussed with an example in Section 3.1. Then, this preliminary notion is refined and recursive learning is fully developed in Section 3.2. Test generation with recursive learning is discussed in Section 3.3.

## 3.1 Example

To give the reader a quick understanding of how recursive learning enumerates to extract necessary assignments, this section introduces a preliminary and simplified recursive learning routine. This routine will be refined in subsequent sections. All reasoning in recursive learning is driven by direct implications. The notion of direct implications has been introduced in Section 2.3. Table 3.1 depicts the preliminary recursive learning routine.

Figure 3.1 shows an example to illustrate *demo_recursive_learning()* of Table 3.1. For the time being, disregard the pads $x_e$, $x_g$, $x_h$, $y_e$, $y_g$ and $y_h$ which will be used later to insert additional circuitry. Consider signals $i_1$ and $j$. Let $i_1 = 0$ and $j = 1$ be two unjustified lines and the reader may verify that $i_2 = 0$ and $k = 1$ are necessary assignments. This is easy to observe as the nodes are labeled in a special way. Signals labeled by the same letter but different subscript always assume the same logic value for an arbitrary combination of value assignments at the inputs; i.e., nodes labeled with the same letter are functionally equivalent. (The pads $x_i$, $y_i$ are excluded from this convention)

It is now explained through a step by step discussion how recursive learning derives the necessary assignments $i_2 = 0$ and $k = 1$ for the given value assignments $i_1 = 0$ and $j = 1$ in Figure 3.1. This is done in Table 3.2, which lists the different steps for unjustified line $i_1 = 0$ when procedure *demo_recursive_learning()* is performed. The first column represents the situation of value assignments before and after recursive learning has been performed. First, direct implications are performed for the initial set of value assignments, $S = \{i_1 = 0, j = 1\}$. In our example, this does not result in any additional value assignment. The algorithm proceeds by examining *justifications* for the current set of unjustified lines $U^0$ which have been created by direct implications. The notion of *justification* will be defined in Section 3.2.1.

The unjustified line $i_1$ can be justified by assigning the controlling value to one of the inputs of gate $i_1$. We assign the controlling value first at $g_1$ (column 2). With the direct implications for $g_1 = 0$ we obtain the unjustified lines $e_1$ and $f_1$ in recursion level $r = 1$. These signals are treated by recursively calling the same procedure again (column 3). For the unjustified line $e_1 = 0$ we examine the justifications $a_1 = 0$ and $b_1 = 0$. In *both* cases we obtain $e_2 = 0$. This value assignment is obviously *necessary* to justify the unjustified line $e_1$. Now we proceed in the same way with unjustified line $f_1$ and obtain that $f_2 = 0$ is necessary. Returning to level 1 the implications can be completed and the necessary assignments $k = 1$ and $i_2 = 0$ are obtained. Note that all signal assignments that have been made during this procedure in each level $r$ have to be erased again as soon as the procedure returns to the previous level $r$-1. This is not explicitly stated in the pseudo code. Only those values that are "learned" to be necessary are transferred to the previous level of recursion.

```
/* this procedure operates on a global data structure representing the gate
  netlist of the circuit with possibly pre-set value assignments at some of the
  nodes, it takes as input a set of new value assignments S in the circuit, the
  parameter r is the current recursion level, initially r = 0   */

demo_recursive_learning(S, r)
{
  make direct implications for value assignments S in the circuit and
  set up a list of resulting unjustified lines U^r;

  if (logic inconsistency occurred)
      return INCONSISTENT;

  for (each unjustified line g ∈ U^r)
  {
      /* examine justifications at unjustified line */
      for (each input i of gate g)
      {
          /* try justifications */
          S = {i := controlling value}
          consistent_i = demo_recursive_learning(S, r+1);
      }

      /* extract necessary assignments */
      for (all signals f in the circuit, such that f assumes the same
            logic value V for all consistent justifications i)
      {
          /* learning */
          assign f := V;
          make direct implications in the circuit;
          if (logic inconsistency occurred)
              return INCONSISTENT;
      }

      /* check logic consistency */
      if (consistent_i = INCONSISTENT for all justifications i)
          return INCONSISTENT;
  }
  return CONSISTENT;
}
```

**Table 3.1:** Routine *demo_recursive_learning()*

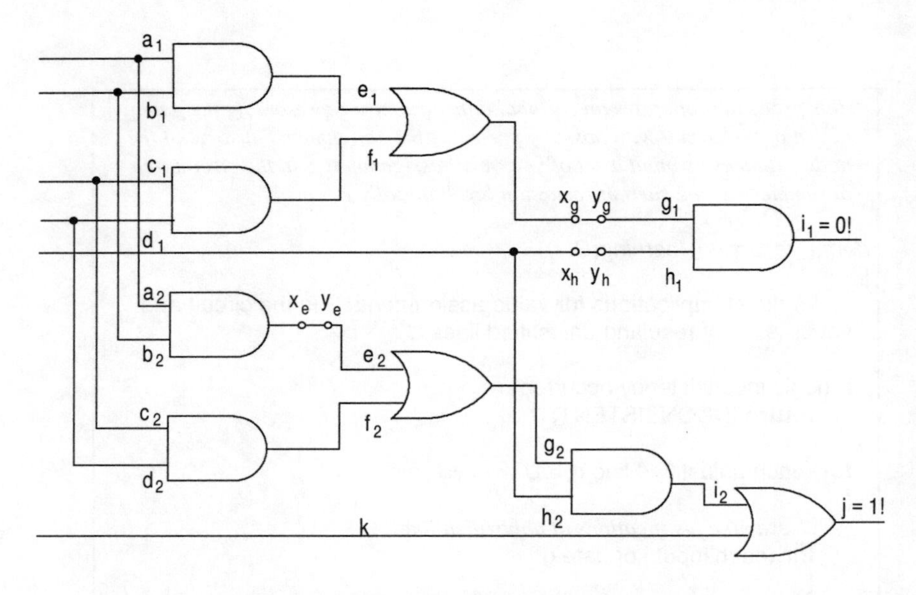


**Figure 3.1:** Circuit example for illustrating *demo_recursive_learning()*

One important aspect of this procedure is that the unjustified lines are considered *separately*. At each unjustified line we examine the different justifications and then move on to the next unjustified line. This raises a natural question as to how we take into account that some necessary assignments result from the presence of *several* unjustified lines, without trying the justification at one unjustified line in combination with the justifications at other unjustified lines. The fact that each unjustified line is considered separately is an important attribute of recursive learning and reflects that the underlying search in recursive learning is based on what we refer to as *AND/OR reasoning*. AND/OR reasoning is fundamentally different from the usual variable enumeration. The AND/OR reasoning in recursive learning will be studied in more detail in Section 4.2.

Note that the necessary assignment $k = 1$ in the above example is due to both unjustified lines $i_1 = 0$ and $j = 1$ and is correctly derived by *demo_recursive_learning()* although the justifications for the two unjustified lines are considered separately. The interdependence of different unjustified lines is taken into account because forward implications are performed, which check the consistency of the current justifications against all other unjustified lines. However, the completeness of forward implications is not always guaranteed and therefore this preliminary version of the recursive learning routine *demo_recursive_learning()* may fail to identify all necessary assignments. In order to understand what extension has to be made in order to identify *all* necessary assignments this aspect is examined more closely: Figure 3.2 shows an example of how forward implications can be incomplete.

| Level $r = 0$ | Level $r = 1$ | Level $r = 2$ |
|---|---|---|
| $i_1 = 0$ (unjust.) | unjust. line $i_1 = 0$: | unjust. line $e_1 = 0$: |
| $j = 1$ (unjust.) | | |
| | 1. justif.: $g_1 = 0$ | 1. justif.: $a_1 = 0$ |
| enter next | $\Rightarrow e_1 = 0$ (unjust.) | => $a_2 = 0$ |
| recursion $\rightarrow$ | $\Rightarrow f_1 = 0$ (unjust.) | => $\mathbf{e_2 = 0}$ |
| $U^0=\{i_1 = 0, j = 1\}$ | | |
| | enter next | 2. justif.: $b_1 = 0$ |
| | recursion $\rightarrow$ | $\Rightarrow b_2 = 0$ |
| | $U^1 = \{e_1 = 0, f_1 = 0\}$ | $\Rightarrow \mathbf{e_2 = 0}$ |
| | $e_2 = 0$ | <====== |
| | | unjust. line $f_1 = 0$: |
| | | 1. justif.: $c_1 = 0$ |
| | | $\Rightarrow c_2 = 0$ |
| | | $\Rightarrow \mathbf{f_2 = 0}$ |
| | | 2. justif.: $d_1 = 0$ |
| | | $\Rightarrow d_2 = 0$ |
| | | $\Rightarrow \mathbf{f_2 = 0}$ |
| | | <====== |
| | $f_2 = 0$ | |
| | $\Rightarrow g_2 = 0$ | |
| | $\Rightarrow \mathbf{i_2 = 0}$ | |
| | $\Rightarrow \mathbf{k = 1}$ | |
| | 2. justif.: $h_1 = 0$ | |
| | $\Rightarrow h_2 = 0$ | |
| | $\Rightarrow \mathbf{i_2 = 0}$ | |
| | $\Rightarrow \mathbf{k = 1}$ | |
| $k = 1$ | <====== | |
| $i_2 = 0$ | unjust. line $j = 1$: | |
| | ... | |

**Table 3.2:** Using *demo_recursive_learning()*

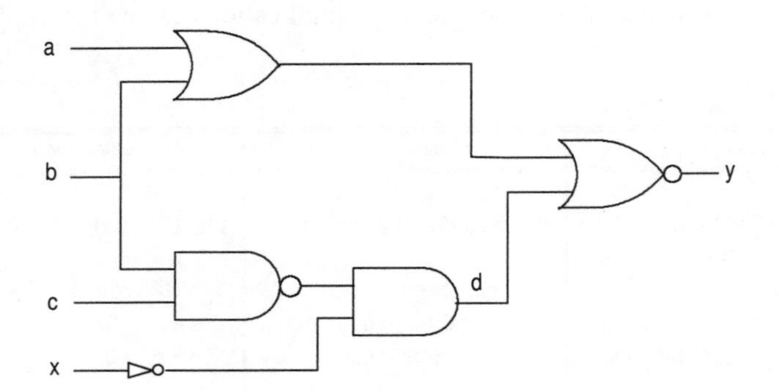


**Figure 3.2:** Example of incomplete direct forward implication

Consider signal $x$ in Figure 3.2. No forward implication can be made for signal $d$ after the assignment $x = 0$ has been made. However, it is easy to see that now both assignments $d = 0$ and $d = 1$ result in the assignment $y = 0$. Hence the forward implication $x = 0 \Rightarrow y = 0$ is true.

In practice, incompleteness of forward implications seems to be a minor problem. When learning is performed for a particular unjustified line and necessary assignments are missed because of incomplete forward implications then there often exists some other unjustified line for which these necessary assignments can be learned.

This can be illustrated with Figure 3.1 and Figure 3.2. If we add the circuitry of Figure 3.2 between the pads $x_e$ and $y_e$ in Figure 3.1, such that signal $x$ of Figure 3.2 is connected to $x_e$ and $y$ is connected to $y_e$, it can be observed that the procedure for unjustified line $i_1$ will no longer yield the necessary assignments $k = 1$ and $i_2 = 0$. However, as the reader may verify, the necessary assignments $i_2 = 0$ and $k = 1$ can still be identified when learning is performed for unjustified line $j$. Experiments show that this is a frequent phenomenon which can be taken into account as explained in Section 3.3.2. Nevertheless, *demo_recursive_learning()* can miss necessary assignments because of incomplete forward implications. The reader may verify, that learning at line $j$ will also fail to identify $k = 1$ and $i_2 = 0$, if we add similar circuitry as in Figure 3.2 (remove the inverter and replace NOR by OR) between the pads $x_g$, $y_g$ and $x_h$, $y_h$. The reason for this incompleteness is that unjustified lines are not the only logic constraints at which learning has to be initialized. In order to overcome this problem the concept of unjustified lines will be generalized in the next section.

## 3.2 Determining all necessary assignments

In a FAN type algorithm, as outlined in Section 2.4, necessary assignments are derived in two ways. The first is based on a structural examination of conditions for fault detection [FuSh83], [KiMe87]. Second, it is the task of an *implication procedure* to derive necessary assignments which result from previously made signal assignments. In this section, methods are developed for identifying all necessary assignments of both types. In Section 3.2.1, a technique is presented that can make *all* implications for a given situation of value assignments. Section 3.2.2 introduces an algorithm for *complete unique sensitization,* i.e., a technique to derive *all* necessary assignments resulting from the requirement to propagate the fault signal to at least one primary output.

The methods proposed in Section 3.2.1 and Section 3.2.2 are based on *AND/OR enumeration* as will be further explained in Section 4.2. Necessary assignments are extracted by "monitoring" the enumeration process. This philosophy is related to the *learning* techniques described in Section 2.7. However, the learning methods proposed in Section 2.7 are based on monitoring a conventional variable enumeration (restricted to one variable) as shown in Table 2.4. The difference between conventional variable enumeration and AND/OR enumeration is the subject of Chapter 4 and will not be further considered here. When monitoring AND/OR enumeration, assignments being necessary in certain subspaces of this search are extracted by certain rules and used in subsequent operations to identify necessary assignments that are valid for the entire search space. This monitoring of AND/OR enumeration is called recursive learning and will be described in detail in the following two sections.

### 3.2.1 The Complete Implication Procedure

We now formulate the basic notions of recursive learning which are also the basis for the AND/OR reasoning graphs to be introduced in Chapter 4. The following definitions use the common notation of a "specified signal" as defined in Section 2.5 for the logic alphabets $B_2$, $B_3$, and $B_5$. For example in Roth's logic alphabet $B_5 = (0, 1, X, D, \bar{D})$, a signal is specified, if it has one of the values 0, 1, D, or $\bar{D}$. It is unspecified if it has the value X. It was pointed out in the previous section that the concept of unjustified lines must be generalized to overcome the incompleteness of the preliminary implication algorithm *demo_recursive_learning()*. Here, the more general concept of *unjustified gates* is introduced.

**Definition 3.1:** Given a gate $g$ that has at least one specified input or output signal and the values at $g$ are logically consistent: *Gate g is called unjustified,* if there are one or more unspecified input or output signals of $g$ for which there exists a combination of value assignments that is logically inconsistent at $g$. Otherwise, $g$ is called justified.

Note that the notion of unjustified or justified gates only applies to gates with at least one specified input or output signal. If no signal is specified the gate is termed *unspecified.* We can now define the notion of *complete* or *precise* implications:

**Definition 3.2:** For a given combinational network, let $S$ be a set of value assignments, $S = \{f_1 = V_1,\ f_2 = V_2,\ \ldots,\ f_k = V_k\}$, where $V_i$ , $i = 1, \ldots k$, is some specified value of $B$, and $B$ is some logic alphabet, $B_2$, $B_3$ or $B_5$. Let $g$ be an arbitrary but unspecified signal in the circuit and $W$ some specified logic value. If all consistent sets of value assignments contain the assignment $g = W$ and form a superset of $S$, such that there are no unjustified gates in the circuit, then the assignment $g = W$ is called *necessary* for the given situation of value assignments $S$. Implications are called *precise* or *complete* when they determine *all* necessary assignments for a given situation of value assignments.

Note that there is a subtle difference between the problem of finding all *necessary* assignments for a given set of value assignments and the problem of identifying all signal values in the circuit that are *uniquely determined.* If a value is uniquely determined *independent* of the given set of value assignments, i.e., the given set of value assignments can be justified without any involvement of this signal, then complete implications as defined above may not identify this value. Complete implications as defined above identify all those uniquely determined values that are a proper consequence of the given set of value assignments.

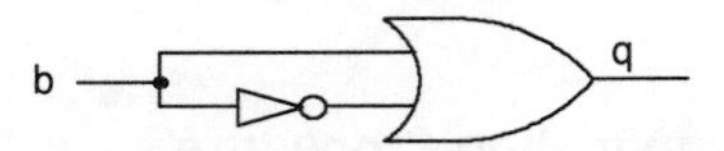


**Figure 3.3:** Shakespeare's circuit example

In practice, this does not cause any problem. Clearly, prior to making some value assignments in the network there can only be uniquely determined values if the corresponding nodes implement constant functions and are therefore redundant. The most simple case for such a situation is shown in Figure 3.3. Of course, such a constant node can also be identified by a complete implication technique if the opposite value is assigned to that node. Then, the constant value is in conflict with the given set of value assignments and the complete implication routine produces a conflict. In spite of the subtle difference, for the rest of this book we will use the terms *necessary assignments* and *uniquely determined values* synonymously.

**Definition 3.3:** Let $f_1, f_2, \ldots f_n$ be some unspecified input- or output signals of an unjustified gate $g$ and let $V_1, V_2, \ldots, V_n$ be logic values which specify them. The set of signal assignments, $J = \{f_1 = V_1, f_2 = V_2, \ldots f_n = V_n\}$, is called *justification* for g, if the combination of value assignments in $J$ makes $g$ justified.

Unjustified gates describe the locations in the circuit *where* values have to be injected during the recursive learning process in order to examine their logical consequences. Justifications determine *what* signal values have to be injected at the unjustified gates. The left column of Figure 3.4 shows examples of unjustified and justified gates. The right column depicts the corresponding justifications.

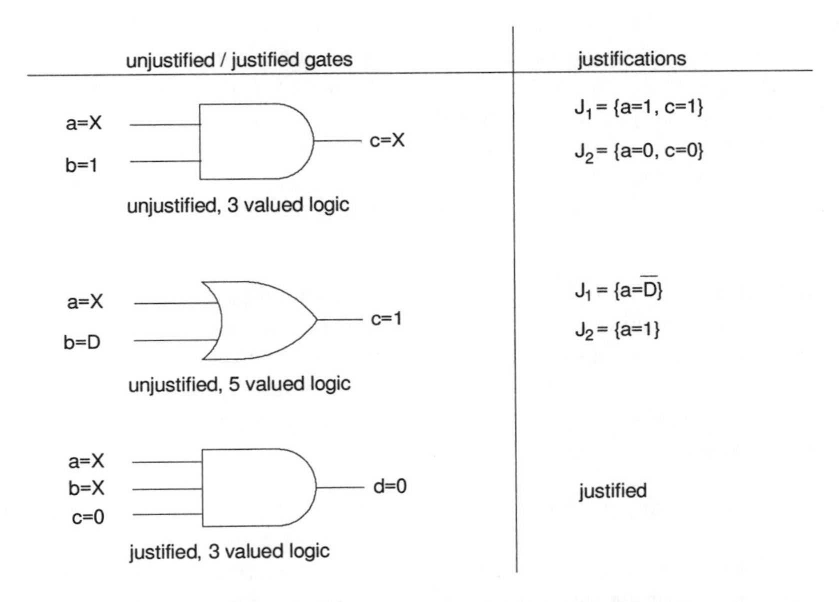


**Figure 3.4:** Unjustified gates and justifications

**Definition 3.4:** Let ${}^gC$ be a set of $m$ justifications $J_1, J_2, \ldots J_m$ for an unjustified gate $g$. If there is at least one justification $J_i \in {}^gC$, i = 1, 2...$m$ for any possible justification $J^*$ of $g$, such that $J_i \subseteq J^*$, then set ${}^gC$ is called *complete*.

For a given unjustified gate, it is straightforward to derive a complete set of justifications. In the worst case, this set consists of all consistent combinations of signal assignments representing a justification of the considered gate. Often though, the set can be smaller, as for the example shown in Figure 3.5.

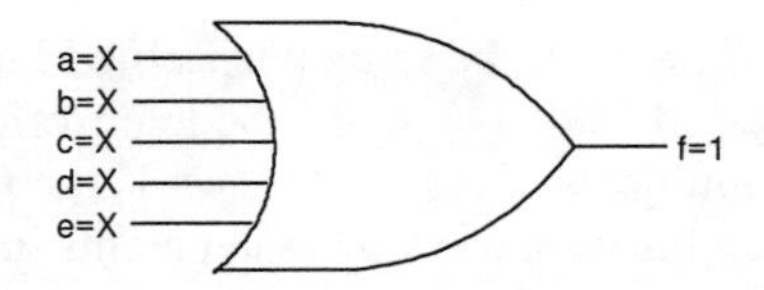


**Figure 3.5:** Complete set of justifications

The following represents a complete set of justifications for the unjustified gate in Figure 3.5: $C = \{J_1, J_2, J_3, J_4, J_5\}$ with $J_1 = \{a = 1\}$, $J_2 = \{b = 1\}$, $J_3 = \{c = 1\}$, $J_4 = \{d = 1\}$, $J_5 = \{e = 1\}$. Note, that for example the justification $J^* = \{a = 1, b = 0\}$ does not have to be in $C$ since all assignments in $J_1$ are contained in $J^*$.

The concept of justifications for unjustified gates is essential toward understanding how learning is used to derive necessary assignments. Assignments are necessary for the justification of a gate, if they have to be made for *all* possible justifications. By definition, all assignments which have to be made for all justifications that represent a complete set of justifications, also have to be made for any other justification at the respective gate. Hence, for a given gate, it is sufficient to consider a complete set of justifications in order to obtain assignments being necessary for all justifications.

> **Definition 3.5:** Let $R(S)$ be the set of value assignments $f_i = V_i$ for those variables $f_i$ in a combinational network whose values have been changed by making implications for a given set of value assignments $S$. Further, $U(R)$ is the set of variable assignments at the outputs of those unjustified gates, which have an input with a variable assignment contained in $R$. The set $E(S) = R(S) \cup U(R)$ is called the *event list E for S*.

In other words, when performing (e.g., direct) implications for a given set of value assignments $S$, the event list $E$ contains all variables whose values have been changed. This includes the output signals of new unjustified gates. Furthermore, the output signals of old unjustified gates are included if their status has changed, i.e., if one of their inputs has assumed a different value.

Table 3.3 depicts procedure *make_all_implications()*, which is able to make complete implications for a given set of value assignments in a combinational network. The shaded regions in Table 3.3 mark the AND/OR enumeration which is the basis of recursive learning and which is further discussed in Section 4.2. The routine is similar to the preliminary recursive learning approach in the previous section. The only difference is that it uses the more refined notions of injecting justifications as elements of a complete set of justifications at unjustified gates in the circuit.

```
/* this procedure operates on a global data structure representing the
   gate netlist of the circuit with possibly pre-set value assignments at
   some of the nodes, it takes as input a set of new value assignments S in
   the circuit, the parameter r is the current recursion level, initially r = 0,
   r_max is a user defined aborting criterion   */

make_all_implications(S, r, r_max)
{
   make all direct implications for S in circuit and
   set up a list U^r of unjustified gates in event list E(S);

   if (value assignments are logically inconsistent)
      return INCONSISTENT;

   if (r<r_max)
   {
      for (each unjustified gate g in U^r)
      {
         set up list of justifications ^gC^r;

         /* try justifications */
         for (each justification J_i ∈ ^gC^r)
            consistent_i := make_all_implications(J_i, r+1, r_max);

         /* extract necessary assignments */
         for (all signals f in the circuit, such that f assumes the
         same logic value V for all consistent justifications i)
         {   /* learning */
            assign in the circuit, f := V;
            make all direct implications in circuit and update U^r;
            if (value assignments are logically inconsistent)
               return INCONSISTENT;
         }
         /* check logic consistency */
         if (consistent_i = INCONSISTENT for all i)
            return INCONSISTENT;
      }
   }
   return CONSISTENT;
}
```

**Table 3.3:** Routine *make_all_implications()*

The reasoning in *make_all_implications()* as shown in Table 3.3 is based on direct implications for injected sets of value assignments that represent justifications for unjustified gates. Instead of evaluating the logical consequences of these justifications by *direct* implications it may be beneficial to use other basic implication techniques, e.g., *transitive closure* [ChAg91], [ChAg93], [AgBu96] or the implication graph of [HeWi95], at the core of *make_all_implications()*.

**Theorem 3.1:** The procedure in Table 3.3 makes complete implications; i.e., a finite $r_{final}$ always exists such that *make_all_implications(0,* $r_{final}$*)* determines all necessary assignments for a given set of value assignments with unjustified gates, $U^0$. If the initial set of value assignments is impossible in the circuit, then a logic inconsistency is produced (see Appendix for a proof).

**Example 3.1:** Figure 3.6 shows a combinational circuit to illustrate *make_all_implications()*. Table 3.4 lists the single steps to perform the implication $p = 1 \Rightarrow q = 1$. Note that the learning techniques [ScTr87], [ScTr88] and [ScAu89] cannot perform this implication.

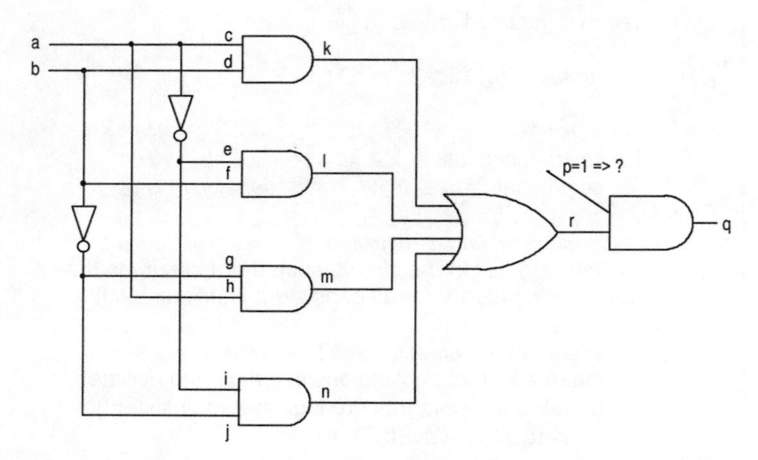


**Figure 3.6:** Example circuitry to illustrate *make_all_implications()*

Figure 3.7 depicts a scheme for a better understanding of the general procedure of recursive learning and Theorem 3.1. Consider *make_all_implications()* as part of a test generation process as described in Section 2.4. During the test generation process, optional assignments are made. After each optional assignment, the resulting necessary assignments must be determined. This is the task of the implication procedure. Many necessary assignments can be determined by performing direct implications only. Direct implications handle the special case where there is only one possible justification for an unjustified gate. Note that this represents another possibility to define "direct" implications. The circles in Figure 3.7 stand for sets of value assignments in the circuit. The left column in Figure 3.7 shows the situation after an optional assignment during the test generation process. After direct implications a situation of value assignments results where only those unjustified gates (dark spots in Figure 3.7) remain that allow for more than one justification. These are examined by learning.

| Level $r = 0$ | Level $r = 1$ | Level $r = 2$ |
|---|---|---|
| (generally valid signal values) | for unjust. gate $q$: | for unjust. gate $k$: |
| $p = 1$ (unjust.)<br>enter<br>next level → | 1. justif.: $q = 0$, $r = 0$<br>⇒ $k = 0$ (unjust.)<br>⇒ $l = 0$ (unjust.)<br>⇒ $m = 0$ (unjust.)<br>⇒ $n = 0$ (unjust.)<br>enter next<br>level → | 1. justification $c = 0$<br>⇒ $e = 1$<br>⇒ $f = 0$ (since $l = 0$)<br>⇒ $i = 1$<br>⇒ $j = 0$ (since $n = 0$)<br>⇒ inconsistency<br>at $b$<br><br>2. justification $d = 0$<br>⇒ $g = 1$<br>⇒ $h = 0$ (since $m = 0$)<br>⇒ $j = 1$<br>⇒ $i = 0$ (since $n = 0$)<br>⇒ inconsistency<br>at $a$<br><br>current situation of value assignments inconsistent |
| | 1. justif. inconsistent | <====== |
| | 2. justif.: $q = 1$, $r = 1$<br>$r = 1$ (unjust.)<br>enter<br>next level → | for unjust. gate $r$:<br><br>1. justification $k = 1$<br>⇒ ...<br>2. justification $l = 1$<br>⇒ ...<br>3. justification $m = 1$<br>=> ...<br>4. justification $n = 1$<br>⇒ ...<br>(no new information learned)<br><====== |
| $q = 1$<br>$r = 1$ | $q = 1$ and $r = 1$ are common for all consistent justifications (there is only one)<br><====== | |

**Table 3.4:** Demonstrating procedure *make_all_implications()*

Recursive learning examines the different justifications for each unjustified gate leading to new situations of value assignments in the first level. If value assignments are valid for all possible justifications of an unjustified gate in level 0; i.e., if they lie in the intersection of the respective sets of value assignments in level 1 (shaded area), then they actually belong to the set of value assignments in level 0. This is indicated schematically in Figure 3.7. However, the sets of value assignments in level 1 may be incomplete because they also contain unjustified gates and the justifications in level 2 have to be examined. This is continued until the maximum recursion depth $r_{max}$ is reached. This immediately leads to the question: how many levels have to be explored in order to perform complete implications? Unfortunately, it is neither possible to predict how many levels of recursion are needed to derive all necessary assignments, nor is it possible to determine if all necessary assignments have been identified after the procedure with a certain recursion depth has been completed. The choice of $r_{max}$ is subject to heuristics and depends on the application for which recursive learning is used. For test generation, an algorithm to choose $r_{max}$ will be presented in Section 3.3.1.

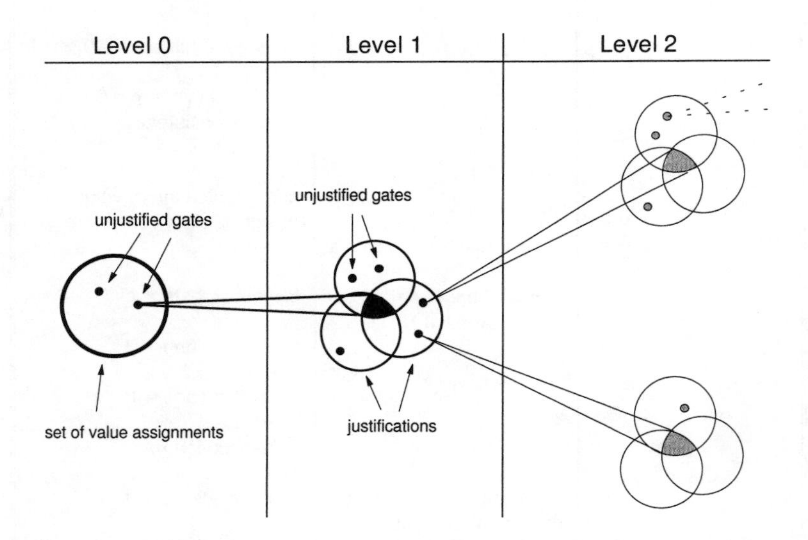


**Figure 3.7:** Schematic illustration of recursive learning

In general, it can be expected that the maximum depth of recursion needed to determine all necessary assignments is relatively low. This can be explained as follows: Note, that value assignments in level $i$+1 are only necessary for level $i$ if they lie within the intersection in level $i$+1. In order to be necessary in level $i$-1 they also have to be in the intersection of level $i$ and so forth. It is important to realize however, that we are only interested in the necessary assignments of level 0. It is not very likely that, e.g., a value necessary in level 10 also lies in the corresponding intersections of levels 9, 8, 7,... 1 and, hence, is not likely to be necessary in level 0. Necessary assignments of level 0 are usually determined by considering only a few levels of recursion. This corresponds to the plausible concept that unknown logic constraints (necessary assignments) must lie in the "logic neighborhood" of the known logic constraints by which they are caused. Intuitively, a lot of recursions are needed only if there is a lot of redundant circuitry. Consider the circuits in Figure 2.12, Figure 3.1 and Figure 3.6: Necessary assignments are only missed by direct implications because the circuits are sub-optimal. In the scheme of Figure 3.7 the intersections of justifications (shown as shaded areas) indicate logic sub-optimality in the circuit. In fact, making *indirect* implications and identifying sub-optimal circuitry seem to be closely related. This observation is very important and represents the basis for the multi-level optimization approach described in Section 5.6.

We use Figure 3.7 to understand the proof of Theorem 3.1. Notice that the process of making all implications terminates, even if the parameter $r_{max}$ in *make_all_implications(r, $r_{max}$)* is chosen to be infinite. At some point the justifications must reach primary inputs and outputs so that no new unjustified gates requiring further recursions will be there. In Figure 3.7 such justifications are represented by circles that do not contain dark spots. If the individual justifications for a considered unjustified gate do not contain unjustified gates, it is impossible (because of Definition 3.1) that these sets of value assignments produce a conflict with justifications of some other unjustified gates. Since a complete set of justifications is examined and the same argument applies to every unjustified gate in the previous recursion level, it is guaranteed that *all* necessary assignments for the previous recursion level are identified. This is used in Step 1 of the induction for Theorem 3.1. If all necessary assignments are known in a given recursion level, the intersections of the complete sets of justifications yield all necessary assignments for the previous recursion level and with the preliminary remarks in the proof of Theorem 3.1 the induction is straightforward.

### 3.2.2 The Complete Unique Sensitization Procedure

In principle, the problem of test generation is solved with a precise implication technique as given in Section 3.2.1. Observability constraints can always be expressed in terms of unjustified lines by means of Boolean difference. However, most ATPG algorithms use the concept of a *D-frontier* as introduced in Section 2.4. This makes it easier to consider topological properties of the circuit [FuSh83].

Table 3.5 describes an algorithm to identify all necessary assignments due to the requirement of propagating the fault signal to at least one primary output. In analogy to the previous section where justifications are injected at unjustified gates in order to perform complete implications, this section shows how all conditions for fault propagation can be derived by injecting *sensitizations* (see Section 2.5) at the D-frontier. The D-frontier in a recursion level $r$ shall be denoted $F^r$ and according to Definition 2.5 consists of all newly implied signals having a faulty value and an unspecified successor. Figure 3.8 shows a circuit example with a D-frontier. After fault injection for signal $a$, stuck-at-0, we obtain $F^0 = \{b, e, g, h\}$.

Procedure *complete_unique_sensitization()* (Table 3.5) calls *make_all_implications()* and correctly identifies all assignments necessary to sensitize *at least* one path from the fault location to an arbitrary output. Note, that we are *not* considering only single path sensitization. Along every path sensitized in procedure *complete_unique_sensitization()*, gates become unjustified if there is more than one possibility to sensitize them. This is demonstrated in Table 3.6 for gate $s$ in Figure 3.8.

> **Example 3.2:** Table 3.5 lists procedure *complete_unique_sensitization()*. Table 3.6 is a step by step explanation how *complete_unique_sensitization()* determines the necessary assignment $n = 0$ for the example in Figure 3.8.

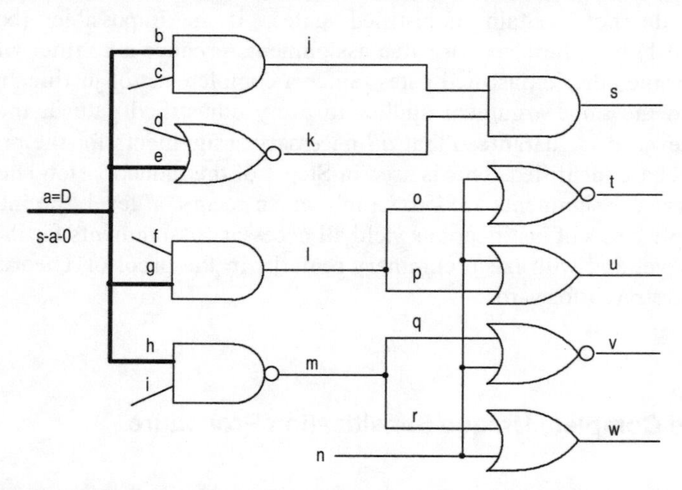


**Figure 3.8:** Circuit to demonstrate *complete_unique_sensitization()*

Procedure *complete_unique_sensitization()* as given in Table 3.5 does not show how to handle XOR gates. However, the extension to XOR gates is straightforward. XOR gates as well as XNOR gates always allow for more than one way to sensitize them. Therefore, the fault propagation has to stop there and the different possibilities to propagate the fault signal have to be tried following the usual scheme for unjustified gates.

```
/*  this procedure operates on a global data structure representing the gate netlist of
    the circuit with possibly pre-set value assignments at some of the nodes, it takes as
    input a D-Frontier F in the circuit, the parameter r is the current recursion level,
    initially r = 0, r_max is a user defined aborting criterion */

complete_unique_sensitization(F^r, r, r_max)
{
    /* sensitizations: for every element in the D-frontier: */
    for (all signals f_i ∈ F^r)
    {
        successor_signal := f_i;

        /* propagate fault signal along adjacent path towards the primary outputs */
        S := ∅;
        while (successor_signal has exactly one successor)
        {
            fault_value := value of successor_signal;
            successor_signal := successor of successor_signal;
            if (successor_signal is output of inverting gate)
                fault_value := INV(fault_value)
            S := S ∪ {successor_signal = fault_value};
        }
        consistent_i := make_all_implications(S, r+1, r_max);
        set up list of new D- frontier F^(r+1);
        if (X-path check for successor_signal fails)
            consistent_i := 0;
        if (r < r_max  and consistent_i = 1)
            consistent_i := complete_unique_sensitization(F^(r+1), r, r_max);
    }
    /* extract necessary assignments: */
    for (all signals f in the circuit, such that f assumes the same
            logic value V for all consistent sensitizations i)
    {   /* learning */
        assign in the circuit, f := V;
        make direct implications in circuit and update U^r and F^r;
        if (value assignments are logically inconsistent or X-path check fails)
            return INCONSISTENT;
    }
    /* check logic consistency: */
    if (consistent_i = INCONSISTENT for all sensitizations i)
        return INCONSISTENT;
    return CONSISTENT;
}
```

**Table 3.5:** Routine *complete_unique_sensitization()*

| Level $r = 0$ | Level $r = 1$ | Level $r = 2$ |
|---|---|---|
| (generally valid signal values)<br><br>$F^0 = \{b, e, g, h\}$<br>enter<br>next level → | D- frontier signal $b$:<br>1. sensitization:<br>successor of $b$:<br>$\Rightarrow j = \mathrm{D}, c = 1$<br>successor of $j$:<br>$\Rightarrow s = \mathrm{D}$, (unjust.)<br>enter next level → | for unjust. gate $s$:<br>1. justification $k = 1$<br>$\Rightarrow$ inconsistent<br>with $e = \mathrm{D}$<br>2. justification $k = \mathrm{D}$<br>$\Rightarrow$ inconsistent<br>with $e = \mathrm{D}$ |
| | 1. sensitization failed | <====== |
| | D- frontier signal $e$:<br>2. sensitization:<br>successor of $e$:<br>$\Rightarrow k = \bar{\mathrm{D}}, d = 0$<br>successor of $k$:<br>$\Rightarrow s = \bar{\mathrm{D}}$ (unjust.)<br>enter next level → | for unjust. gate $s$:<br>1. justification $j = 1$<br>$\Rightarrow$ inconsistent<br>with $b = \mathrm{D}$<br><br>2. justification $j = \bar{\mathrm{D}}$<br>$\Rightarrow$ inconsistent<br>with $b = \mathrm{D}$ |
| | 2. sensitization failed | <====== |
| | D- frontier signal $g$:<br>3. sensitization:<br>successor of $g$<br>$\Rightarrow l = \mathrm{D}, f = 1$<br>several successors<br>of $l$:<br>$\Rightarrow F^1 = \{o, p\}$<br>enter next level → | D- frontier signal $o$:<br>1. sensitization:<br>successor of $o$:<br>$\Rightarrow t = \bar{\mathrm{D}}$, $\boldsymbol{n = 0}$<br><br>D- frontier signal $p$:<br>2. sensitization:<br>successor of $p$:<br>$\Rightarrow u = \mathrm{D}$, $\boldsymbol{n = 0}$ |
| | $\boldsymbol{n = 0}$ | <====== |
| | D- frontier signal $h$:<br><br>4. sensitization:<br>successor of $h$<br>$\Rightarrow m = \bar{\mathrm{D}}, i = 1$<br>several successors<br>of $m$:<br>$\Rightarrow F^1 = \{q, r\}$<br>enter next<br>level → | D- frontier signal $q$:<br><br>1. sensitization:<br>successor of $q$:<br>$\Rightarrow v = \mathrm{D}$, $\boldsymbol{n = 0}$<br><br>D- frontier signal $r$:<br><br>2. sensitization:<br>successor of $r$:<br>$\Rightarrow w = \bar{\mathrm{D}}$<br>$\boldsymbol{n = 0}$ |
| $n = 0$ | <====== $\boldsymbol{n = 0}$ | <====== |

**Table 3.6:** Demonstrating *complete_unique_sensitization()*

## 3.3 Test Generation with Recursive Learning

After describing the basic notions of recursive learning this section discusses an application. Given a complete method to identify all necessary assignments for detecting single stuck-at faults we are ready to develop a test generation algorithm. There are many possibilities to design a test generation algorithm based on recursive learning. There is unlimited freedom to make optional assignments. We are not bound to the strict scheme of the decision tree in order to guarantee the completeness of the algorithm. It is possible to "jump around" in the search space as we wish. Note that this allows attractive possibilities for new heuristics. In order to guarantee completeness, we only have to ensure that the maximum recursion depth for implications and unique sensitization is eventually incremented. Of course, it is wise to keep the maximum recursion depth $r_{max}$ as small as possible to spend minimum time in recursive learning. Only if the precision is not sufficient to avoid wrong decisions, it is sensible to increment $r_{max}$.

### 3.3.1 An Algorithm to Choose the Maximum Recursion Depth $r_{max}$

There are many possibilities in choosing a value for $r_{max}$. In order to examine the performance of recursive learning, we combine it with the FAN algorithm and use the following strategy to generate test vectors: the algorithm proceeds like FAN and makes optional assignments in the usual way. The general procedure for FAN-type test generation has been outlined in Section 2.4. In the same way as for the decision tree, all optional assignments are stored in a stack. Whenever a conflict is encountered we proceed as shown in Figure 3.9. By a conflict we mean that the previous decisions have either led to an inconsistent situation of value assignments or left no possible propagation path for the fault signal (X-path-check failed). The idea behind the routine in Figure 3.9 is that we use recursive learning to leave the non-solution areas as quickly as possible. After a conflict has occurred the previous decision is erased, i.e., the signal at the top of the stack is removed and its value is assigned to X. Now the resulting situation of value assignments is examined with increased recursion depth. If this leads to a new conflict, another decision has to be erased. If there is no conflict this can mean two things: Either the current precision $r_{max}$ is not sufficient to detect that there is still a conflict or we have actually returned into the solution area of the search space. Therefore, it is checked if the opposite of the previous (wrong) assignment is one of the assignments that have been learned to be necessary. This is a good heuristic criterion to determine if the precision has to be increased any further. It also makes sure that we can never enter the same non-solution area twice. The algorithm in Figure 3.9 guarantees the completeness of test generation and redundancy identification without the use of a decision tree.

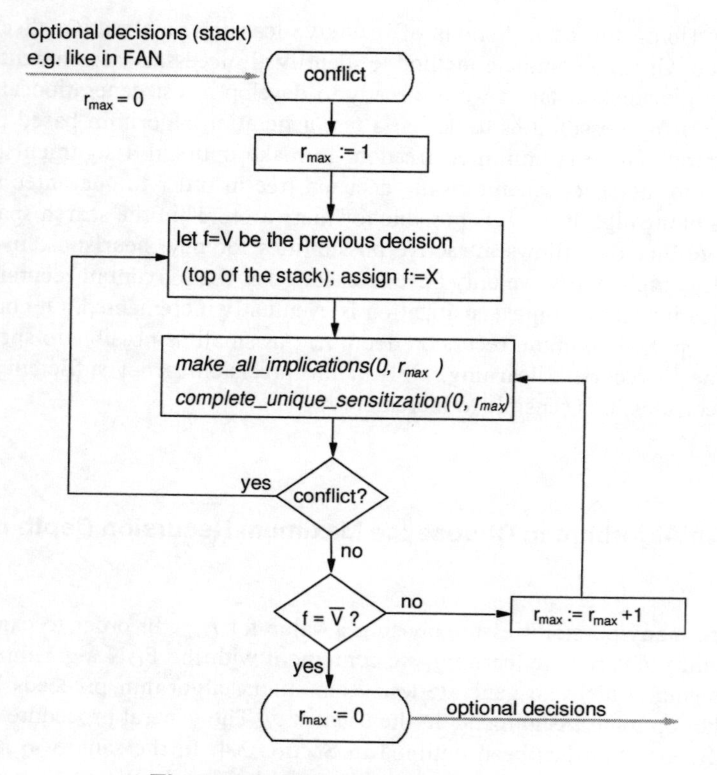


**Figure 3.9:** Algorithm to choose $r_{max}$

The procedure in Figure 3.9 is only one of many possibilities to integrate recursive learning into existing test generation tools. This algorithm allows a fair comparison of recursive learning with the decision tree. With the algorithm of Figure 3.9 we initially enter exactly the same non-solution areas of the search space as with the original FAN algorithm. The point of comparison is how fast the non-solution areas are left by conventional backtracking and by recursive learning.

One disadvantage of the above procedure is that in some cases of redundant faults the algorithm may initially penetrate very deep into non-solution areas and recursive learning has to be repeated many times until all optional value assignments (they are all wrong) are erased step by step. Our current implementation therefore makes use of the following intermediate step (not shown in Figure 3.9 for reasons of clarity): when the algorithm of Figure 3.9 reaches the point where the maximum depth of recursion has to be incremented, we perform recursive learning with incremented recursion depth first only to the situation of value assignments that result if all optional value assignments are removed. If a conflict is encountered, the fault is redundant and test generation is finished. Otherwise, it proceeds as shown in Figure 3.9, i.e., recursive learning is performed with optional value assignments as given on the stack.

### 3.3.2 Experimental Results

In order to examine the performance of recursive learning we use the FAN algorithm. For comparison, we use the original FAN algorithm with the decision tree and a modified version, where we replaced the decision tree by recursive learning. *No additional techniques were used.*

There are two general aspects of recursive learning in a FAN-based environment which we use for better efficiency: first, as discussed in Section 3.1, there are very few cases in practice, where it is necessary to perform learning at unjustified gates with an unspecified output signal. Therefore, learning for unjustified gates with an unspecified output signal is done with a maximum recursion level of $r_{max}-5$ if the current maximum recursion level $r_{max}$ is greater than 5. Otherwise, no learning is performed for such gates. '5' was chosen intuitively to suppress the unnecessary recursions so that they contribute only little to the total CPU-time but still guarantee the completeness of the algorithm.

In order to illustrate the different natures of the two searching schemes, we first compare recursive learning for the traditional search by only considering redundant faults. In our first experiment, we only target all redundant faults in the ISCAS-85 [Brgl85] benchmarks and the seven largest ISCAS-89 [Brgl89] benchmarks.

The results are given in Table 3.7. The first column lists the circuit under consideration and the second column shows the number of the known redundant faults in the circuit. Only these faults are targeted by the test generation algorithm. First, we run FAN with a backtrack limit of 1000, i.e., the traditional searching scheme is used and the fault is aborted after 1000 backtracks. The third column shows the number of backtracks for each circuit. The next two columns show the CPU-time in seconds and the number of aborted faults. In the second run, we use recursive learning instead of the decision tree. Columns 6 and 7 give the CPU-time and the number of aborted faults for recursive learning. The next 4 columns show the number of faults for which the shown high recursion levels were chosen by the algorithm of Figure 3.9. For example for circuit c432, one redundancy could be identified without any learning. Three redundancies were identified in the first recursion level.

| **Results if only redundant faults are targeted** | | FAN with DECISION TREE (bt. limit of 1000) | | | FAN with RECURSIVE LEARNING recursion levels: | | | | | | |
|---|---|---|---|---|---|---|---|---|---|---|---|
| circuit | no. faults targeted | no. of backtracks | **time [s]** | **ab.** | **time [s]** | **ab.** | r0 | r1 | r2 | r3 | r4 |
| c432 | 4 | 3000 | **12** | **3** | **0.2** | **0** | 1 | 3 | - | - | - |
| c499 | 8 | 0 | **0.1** | **0** | **0.1** | **0** | 8 | - | - | - | - |
| c880 | 0 | - | **-** | **-** | **-** | **-** | - | - | - | - | - |
| c1355 | 8 | 0 | **0.1** | **0** | **0.1** | **0** | 8 | - | - | - | - |
| c1908 | 9 | 226 | **5** | **0** | **0.2** | **0** | 7 | - | - | - | - |
| c2670 | 117 | 18862 | **207** | **15** | **112** | **0** | 81 | 25 | 0 | 4 | 7 |
| c3540 | 137 | 5000 | **339** | **5** | **2** | **0** | 132 | 5 | - | - | - |
| c5315 | 59 | 0 | **0.9** | **0** | **0.9** | **0** | 59 | - | - | - | - |
| c6288 | 34 | 0 | **2** | **0** | **2** | **0** | 34 | - | - | - | - |
| c7552 | 131 | 64733 | **3858** | **64** | **36** | **0** | 65 | 66 | - | - | - |
| s5378 | 39 | 0 | **1** | **0** | **1** | **0** | 39 | - | - | - | - |
| s9234 | 443 | 55396 | **4235** | **35** | **75** | **0** | 305 | 106 | 32 | - | - |
| s13207 | 149 | 7159 | **1118** | **1** | **23** | **0** | 131 | 17 | 1 | - | - |
| s15850 | 384 | 2459 | **367** | **2** | **31** | **0** | 360 | 24 | - | - | - |
| s35932 | 3728 | 0 | **837** | **0** | **837** | **0** | 3728 | - | - | - | - |
| s38417 | 161 | 4056 | **1207** | **4** | **47** | **0** | 153 | 8 | - | - | - |
| s38584 | 1345 | 8137 | **3277** | **2** | **227** | **0** | 1300 | 37 | 8 | - | - |

**Table 3.7:** Experimental results for redundant faults (Sun Sparc 5)

The results show the superiority of recursive learning for redundancy identification when compared to the decision tree. Consider the circuit c3540: With the decision tree 5 faults are aborted after performing 1000 backtracks each. There is a total of

5000 backtracks for this circuit. Obviously, for 132 redundancies there have been no backtracks at all. We observe an "all-or-nothing-effect" which is typical for redundancy identification with the decision tree. In the conventional FAN algorithm, if direct implications fail to reveal the conflict, the search space has to be exhausted in order to prove that no solution exists. This is usually intractable.

Recursive learning and the search based on the decision tree have a complementary relationship: the latter is the search for a sufficient solution; its pathological cases are the cases where no solution exists (redundant faults). The former is the search for all necessary conditions. If recursive learning was used to prove that a fault is testable without constructively generating a test vector, the pathological cases are the cases where no conflict occurs, i.e., we have to exhaust the search space if a solution exists (testable faults).

Although recursive learning is always used in combination with making optional decisions to generate a test vector such as given in Figure 3.9, it is not wise to use it in cases where many solutions exist that are easy to find. In those cases it is faster to perform a few backtracks with the decision tree. There are many efficient ways to handle these "easy" faults. In this experiment, we chose to perform 20 backtracks with the decision tree. These are split into two groups of 10 backtracks each. For the first ten backtracks we use the FAN-algorithm with its usual heuristics. When ten backtracks have been performed the fault is aborted and re-targeted. This time we use *orthogonal heuristics*. This means that we always assign the opposite value at the fanout objectives [FuSh83] of what FAN's multiple-backtrace procedure suggests. This time we explore the search space in orthogonal direction compared to our first attempt. For testable faults this procedure has been shown to be very effective. Faults which remain undetected after these 20 backtracks are aborted in phase 1. They represent the difficult cases for most FAN-based ATPG-tools. These pathological faults are the interesting cases when comparing the performance of the two searching techniques.

Table 3.8 shows the results of test generation for the ISCAS-85 benchmarks and for the 7 largest ISCAS-89 benchmarks. After each generated test vector, fault simulation was used to reduce the fault list (fault dropping). No random vectors were used. The first two columns list the circuits under consideration and the number of faults targeted. Columns 3 to 5 show the results of the first phase, in which FAN is performed using its original and orthogonal heuristics with a backtrack limit of ten for each pass. Column 3 gives the number of faults which are identified as redundant and column 4 lists the CPU-times in seconds. CPU-times do not include the time spent on fault simulation. Column 5 gives the figures for the aborted faults. All faults aborted in phase 1 are re-targeted in the second phase, in which we compare the performance of recursive learning to the search based on the decision tree. The meaning of columns 6 to 15 is analogous to Table 3.7.

| Results for collapsed fault list | | 1. PHASE (eliminate easy faults) | | | 2. PHASE for DIFFICULT FAULTS (aborted in 1. phase) | | | | | | | | | |
|---|---|---|---|---|---|---|---|---|---|---|---|---|---|---|
| with fault dropping | | FAN with back-track limit of 10+10 | | | FAN with DECISION TREE (bt. limit of 1000) | | | FAN with RECURSIVE LEARNING | | | | | | |
| | | | | | | | | | | | recursion levels | | | |
| circuit | no. target faults | red | time [s] | ab | red | **time [s]** | **ab** | red | **time [s]** | **ab** | r1 | r2 | r3 | r4 |
| c432 | 93 | 1 | 1 | 3 | 0 | **12** | **3** | 3 | **0.2** | **0** | 3 | - | - | - |
| c499 | 122 | 8 | 4 | 0 | - | **-** | **-** | - | **-** | **-** | - | - | - | - |
| c880 | 95 | 0 | 2 | 0 | - | **-** | **-** | - | **-** | **-** | - | - | - | - |
| c1355 | 185 | 8 | 12 | 0 | - | **-** | **-** | - | **-** | **-** | - | - | - | - |
| c1908 | 178 | 7 | 11 | 2 | 2 | **5** | **0** | 2 | **0.1** | **0** | 2 | - | - | - |
| c2670 | 343 | 98 | 26 | 19 | 8 | **243** | **11** | 19 | **98** | **0** | 8 | 0 | 4 | 7 |
| c3540 | 392 | 127 | 47 | 5 | 0 | **278** | **5** | 5 | **2** | **0** | 5 | - | - | - |
| c5315 | 460 | 59 | 37 | 0 | - | **-** | **-** | - | **-** | **-** | - | - | - | - |
| c6288 | 80 | 34 | 15 | 0 | - | **-** | **-** | - | **-** | **-** | - | - | - | - |
| c7552 | 533 | 67 | 210 | 64 | 0 | **3676** | **64** | 64 | **24** | **0** | 64 | - | - | - |
| s5378 | 460 | 7 | 34 | 0 | - | **-** | **-** | - | **-** | **-** | - | - | - | - |
| s9234 | 1230 | 389 | 267 | 56 | 19 | **3449** | **37** | 54 | **73** | **0** | 22 | 28 | 0 | 6 |
| s13207 | 1096 | 133 | 309 | 16 | 15 | **1070** | **1** | 16 | **16** | **0** | 16 | - | - | - |
| s15850 | 1295 | 374 | 803 | 10 | 8 | **331** | **2** | 10 | **7** | **0** | 10 | - | - | - |
| s35932 | 4794 | 3728 | 1308 | 0 | - | **-** | **-** | - | **-** | **-** | - | - | - | - |
| s38417 | 4021 | 153 | 1108 | 8 | 4 | **1074** | **4** | 8 | **15** | **0** | 8 | - | - | - |
| s38584 | 3301 | 1321 | 890 | 24 | 22 | **3128** | **2** | 24 | **87** | **0** | 16 | 8 | - | - |

**Table 3.8:** Experimental Results for test generation with fault dropping (Sun Sparc 5)

The results show the superiority of recursive learning over the traditional searching method in test generation. There are no aborted faults and the CPU-times for the difficult faults are very short when recursive learning is used. A closer study of the above tables shows that the average CPU-times for each difficult fault is on the same order of magnitude as that for easy faults (with few exceptions). The results show that recursive learning can replace the "whole bag of tricks" which has to be added to the FAN algorithm if full fault coverage is desired for the ISCAS benchmarks. The implementation of our base test generation algorithm is rather rudimentary, so that a lot more speed-up can be gained by a sophisticated implementation. Since the focus of this research has been on the examination of a new searching method (called AND/OR search in Chapter 4) and not on a new test generation tool, no effort was made to combine recursive learning with a selection of state-of-the-art heuristics as they are used, for example, in [WaSh90].

Recursive learning does not affect the speed and memory requirements of ATPG-algorithms as long as it is not invoked; there is no pre-processing or pre-storing of data. This is an important aspect, if test generation is performed under limited resources as Kundu has pointed out [Kund92]. If recursive learning is actually invoked, some additional memory is necessary in order to store the current situation of value assignments in each recursion level. Different flags which steer the implication procedure and store the current signal values at each gate have to be handled separately in each level of recursion. As a rough estimate, this results in an overhead of 25 Bytes for each gate in the circuit if we assume that there are 5 flags to steer the implications and a maximum recursion depth of 5. For a circuit with 100,000 gates we obtain an overhead of 2.5 Mbytes, which is small.

# Chapter 4

# AND/OR REASONING GRAPHS

This chapter develops a general view of the recursive learning procedure of Chapter 3 and presents a generalization of the technique. The main focus is on the basic search process underlying recursive learning. It is shown that the search process in recursive learning is a special instance of an *AND/OR search*. This leads to a *basic reasoning scheme* in Boolean networks based on *AND/OR reasoning graphs*. AND/OR reasoning graphs can identify implications and implicants in multi-level circuits so that basic concepts of two-level circuit theory can be extended and applied to multi-level circuits. This chapter elaborates properties of AND/OR reasoning graphs that are useful in solving design automation problems. Applications will be described in Chapters 5 and 6.

Methods for Boolean reasoning have played an important role in the research of design automation for a long time. Since the work of Quine [Quin52] and McCluskey [McCl56] the notion of an *implicant* is a central element of two-level minimization theory. For an excellent survey of Boolean reasoning for two-level circuits, see [Brow90]. However, for *multi-level* circuits the notion of an implicant has been used only rarely to formulate effective procedures. The reason is that the classical concepts are tailored for the special structure of two-level circuits and do not appropriately take into account the structural properties of general multi-level combinational networks. In this book, we extend the notion of implicants for multi-level circuits and apply this concept to multi-level logic synthesis.

In two-level circuit theory an implicant for a function is a product term such that the function is one if the product term is one. The product term consists of literals that belong to the variables of the Boolean function, i.e., to the inputs of the circuit. An implicant is called prime if no literal can be removed from the product term without destroying its property of being an implicant. The set of all prime implicants for a Boolean function represents a canonical form of the function. In order to make the notion of an implicant useful for describing transformations in multi-level networks a small extension is made. In Section 4.3 we will consider implicants where the literals of the product term may not only belong to the inputs of the circuit but to arbitrary nodes of the combinational network. This contains the classical notion as a special case but allows us to take into account structural properties of multi-level networks.

**Example 4.1:** Consider the node $y$ in the circuit of Figure 4.1. It is immediately obvious that $c$ and $bd$ are prime implicants of function $y$. If we allow that the literals of the implicants do not have to belong exclusively to primary input signals but can belong to arbitrary nodes of the network, additional prime implicants can be determined. Note that $x = 1$ and $z = 0$ can simultaneously occur only for input assignments that produce $y = 1$. Therefore, $x\overline{z}$ is an implicant for $y$. This implicant is prime because neither $x$ nor $\overline{z}$ represents an implicant for $y$.

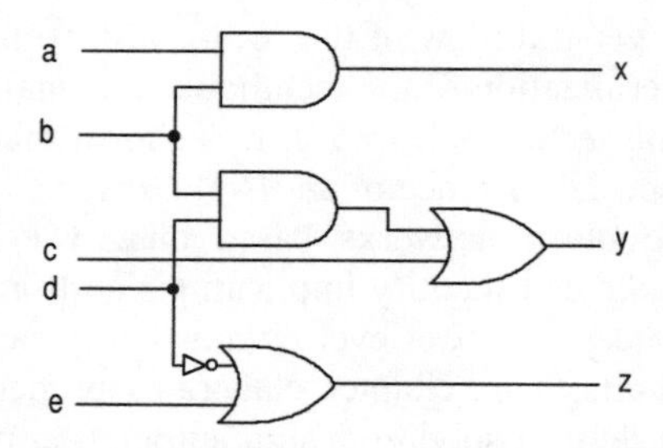


**Figure 4.1:** Circuit example for multi-level implicants

It will be proved in Section 5.6.1 that this extended notion of an implicant can be used to describe arbitrary transformations in a combinational network. Any transformation in multi-level logic synthesis otherwise described by the notions of functional decomposition, kerneling, division, substitution, transduction, etc. (see Chapter 5), can also be described using this extended notion of a network implicant. We address the problem of applying network implicants to multi-level synthesis in Section 5.6. Our present discussion in this chapter is dedicated to a basic reasoning scheme that identifies such implicants and evolves from a generalization of the recursive learning technique presented in the previous chapter.

If $y = 0 \Rightarrow f = 0$ this means that $f = 0$ is necessary for $\overline{y}$ being satisfiable. Using the classical notion of an implicant we could also state that $f$ is an implicant for $y$ since by contraposition it is $f = 1 \Rightarrow y = 1$. We realize that making implications or determining implicants is equivalent to identifying necessary conditions for satisfiability of some Boolean function. Hence, any method that determines implications or calculates implicants must also be a method to solve satisfiability. Therefore, before developing our method to calculate implicants in multi-level networks we first address the satisfiability problem.

The satisfiability problem is the problem to decide whether or not a Boolean function can assume the logic value 1. Many techniques for solving satisfiability in design automation have already been developed. Most of these methods are based on some sort of variable enumeration like decision tree based backtracking or decision diagrams. It is important to note however that these methods for solving satisfiability are hard to use for systematic reasoning. Instead of finding necessary conditions for the satisfiability of a function these methods are directed to produce sufficient solutions. In order to derive implications and implicants the methods presented here, therefore, solve satisfiability in a very different way. Whereas the previous methods enumerate the finite variable space our techniques are related to resolution methods for two-level logic descriptions [Robi65]. However, they operate directly on the multi-level gate netlist description of the circuit. In order to understand how our methods derive implications in multi-level networks it is crucial to realize that satisfiability can be solved by a searching scheme that is radically different from the conventional ones.

## 4.1 OR Search versus AND/OR Search

Every search process can be viewed as a traversal of a directed graph. Standard literature, e.g., [Rich83] distinguishes between two basic types of search graphs. In the simpler case to be considered, the graph is a so-called *OR graph*. A node in the OR graph represents a given problem to be solved and each arc emerging from this node represents a possible move or decision that can be made at the current state of the search process. A solution is found by traversing the graph following certain strategies guided by some heuristics exploiting problem-specific knowledge. Typical strategies are known as *depth-first* search, *breadth-first* search or *best-first* search.

As is well-known however, for some problems it is useful to allow graphs with two types of nodes, AND nodes and OR nodes, that represent a different type of search process. If at a given search state a certain move is made this may lead to *several* new problems that *all* have to be solved. Such AND/OR graphs are the basis for many search methods employed in the field of automatic theorem proving with predicate logic and are used in proof-by-refutation strategies. For a description of general problem-solving techniques in computer science and for more information on the basic concepts of OR graphs and AND/OR graphs, the reader may refer to a standard text book [Rich83].

Also in the field of switching theory, many problems are solved by search techniques. Conventionally, when exploring the implementation of a given Boolean function as a combinational circuit, algorithms enumerate through the finite Boolean space defined by the set of all combinations of value assignments for the input variables. A common search scheme to solve satisfiability or related problems like test generation is the decision tree (see Section 2.2). A decision tree can be considered as an example of an *OR graph*.

Notice that there can also be a different interpretation to such graphs. They not only represent a possible search process to solve a specific problem but, for certain problem formulations (like satisfiability), can also represent the Boolean function of the considered circuit. If exhaustive simulation is represented as tree we obtain a Shannon tree. This tree can be reduced by sharing isomorphic subtrees so that we obtain a *binary decision diagram* (see Chapter 1).

The following observation is relevant to the motivation of this chapter: the conventional concepts to solve satisfiability and related problems in computer-aided circuit design like decision tree based backtracking, exhaustive simulation and Shannon trees or binary decision diagrams can be interpreted as OR trees and not as AND/OR trees. To better understand this interesting point the difference between OR trees and AND/OR trees is illustrated by the following example.

**Example 4.2:** Let us consider Robinson Crusoe's situation after he was shipwrecked and washed ashore a small island. Robinson analyzes his situation and starts thinking how he can leave the island again. We will now follow his reasoning and show how this leads to the AND/OR tree depicted in Figure 4.2.

Robinson can see only two possibilities to leave the island: either he waits for a ship to pass by and pick him up or he builds a boat himself. From the root node representing the assumption that Robinson will leave the island, there are two arcs each leading to a node representing one of the two possibilities. Since at least one of the two possible events must occur in order to bring him home, the root node is an OR node. If he decides to build a boat, he finds, that is only possible if he can get hold of some wood and appropriate tools. Since either prerequisite is not sufficient by itself for the construction of a vessel, the node "build a boat" in the AND/OR tree is an AND node with two successors "wood on the island" and "tools". (AND nodes are marked by an arc.) Of course, he can also wait for a ship to come by some day. In that case, he will need food for the wait and some means of attracting attention when the ship arrives, for example a fire to send a smoke signal. In the AND/OR tree, these necessary conditions are depicted as successors to the AND node "wait for a ship". Also, decent food is not easy to get on a forlorn island. He sees two possibilities, trade with natives or go hunting, and hopes that at least one of them can be realized. Therefore, node "food" is an OR node in the AND/OR tree.

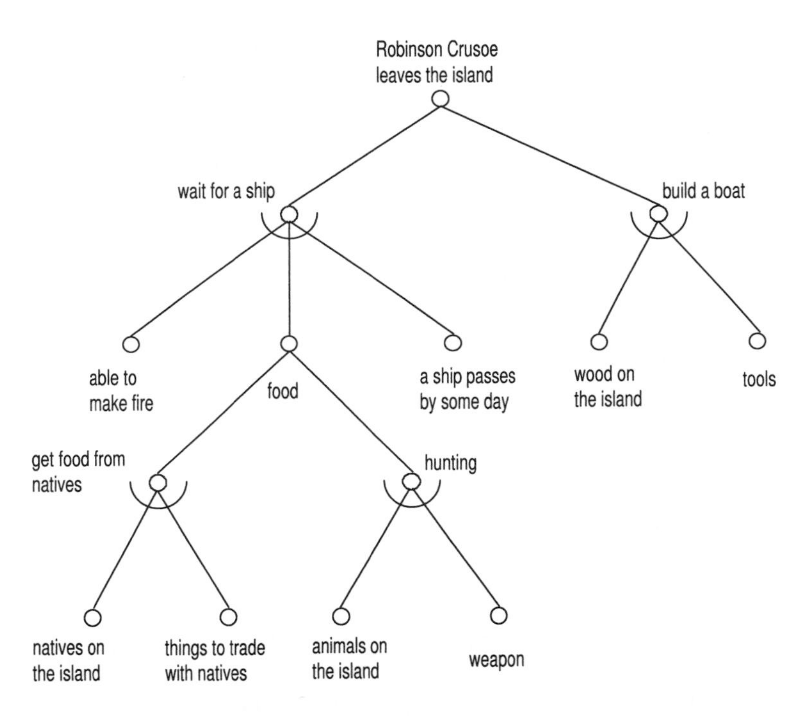


**Figure 4.2:** Example of an AND/OR tree

Robinson has analyzed his situation by determining the necessary conditions for leaving the island. As can be seen, whether he will actually succeed depends on a number of "variables". In principle, each of these "variables" can assume the values "yes" or "no" depending on how lucky he is and what kind of island he is on. Robinson has determined what "values" of these "variables" would be necessary for different scenarios of leaving the island. Since these requirements do not contradict each other the situation is not hopeless and he has a chance of leaving.

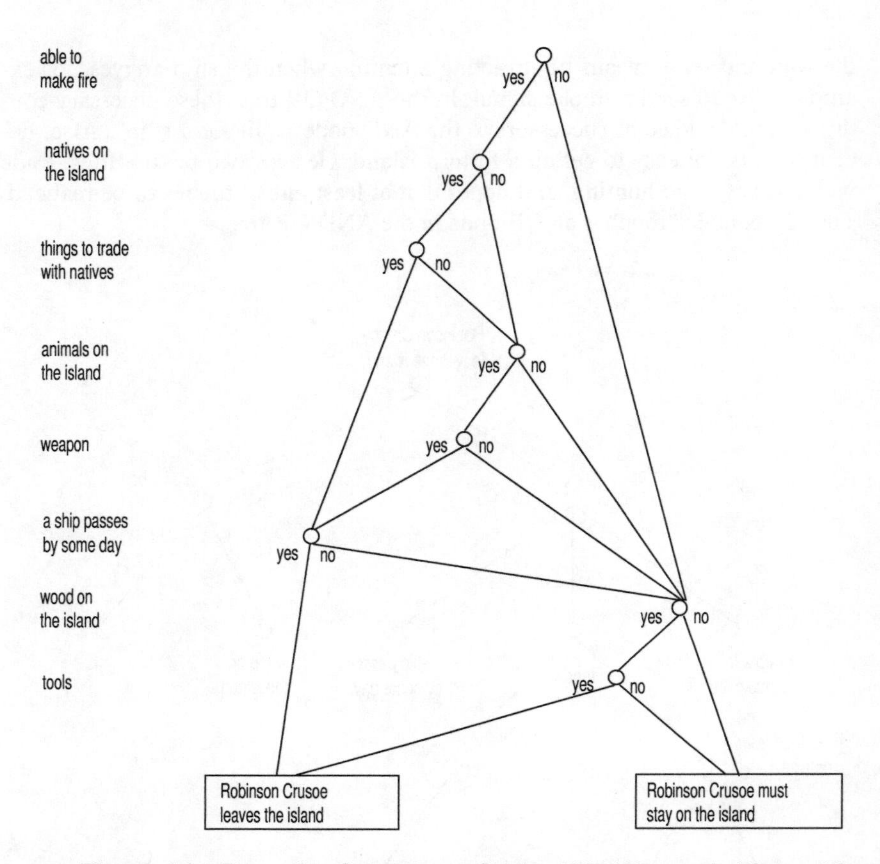


**Figure 4.3:** Example of an OR tree for the AND/OR tree in Figure 4.2

Of course, in order to decide whether or not he can leave it is also possible to check the "variables" one after the other. This leads to representing the problem as an OR tree. In Figure 4.3 the corresponding OR tree is depicted. There are several ways to build an OR tree for the AND/OR tree of Figure 4.2. The tree shown in Figure 4.3 is built in analogy to building OBDDs for Boolean functions.

Note that the OR nodes in the OR tree of Figure 4.3 correspond to the OR leaves of the AND/OR tree of Figure 4.2. The order in which the "variables" appear has a strong influence on the size of the tree. It is interesting to observe that, at least in this example, the structure of the AND/OR tree suggests a good order. The OR nodes in the OR tree follow the same order in which the leaves of the AND/OR tree appear from the left to the right. This suggests that analyzing AND/OR trees for Boolean functions may also lead to useful methods for OBDD variable ordering. However, in this book, this aspect is not further considered.

There is another reason why AND/OR trees are of interest to us in the field of design automation. They turn out to be more suitable for *systematic reasoning* than OR trees. For example, from the AND/OR tree of Figure 4.2 the following implication can be derived in a simple way:

(no natives on the island) and (no weapon) and (no tools)
$\Rightarrow$ (must stay on the island)

In Section 4.3 it will be explained in more detail how such implications can be extracted from an AND/OR tree. In principle, this implication can also be derived from the OR tree in Figure 4.3. Note that all paths from the root node to one of the terminal nodes that pass through the "no" branches of "natives on the island", "weapon" and "tools" lead to the terminal node "Robinson Crusoe must stay on the island". This may be easy to see in this small example but for larger trees such an analysis becomes intractable. The problem is that we cannot exclusively consider these three nodes. We also have to go through nodes "able to make fire", "animals on the island" and "wood on the island". In fact, it does not matter whether there are animals on the island if Robinson has no weapon. This information is obvious in the AND/OR tree, however, moving from the top to the bottom in the OR tree we cannot skip the node "animals on the island" although this node does not have any influence on the result if Robinson does not have a weapon.

Note that any Boolean expression can be understood as an AND/OR tree. However, such a general AND/OR tree does not decide whether the implemented Boolean function is satisfiable. As mentioned, this problem is usually solved by resorting to an OR tree based enumeration. Here we are interested in specific AND/OR trees for Boolean functions that decide satisfiability. Solving satisfiability by AND/OR search requires a totally different way of stepping through the circuit and its variables compared to conventional backtracking methods.

The differences between the two searching schemes are of great practical interest in the field of design automation. As illustrated in the above example OR search techniques are hard to use for *systematic reasoning*. Specifically, for some Boolean statement $A$ we would like to derive some statement $B$ that is true if $A$ is true, i.e., $A \Rightarrow B$. Previous representations of Boolean functions are not well suited for this kind of task. For example, given a statement $A$, a BDD-based approach cannot *derive* or *imply* statement $B$, it can only *check* if $A \Rightarrow B$ is true when both $A$ and $B$ are given. By way of contrast, as will be shown in this section, AND/OR reasoning techniques can determine implications and implicants in multi-level networks. This is of great practical relevance as will be shown in Chapters 5 and 6.

## 4.2 AND/OR Reasoning Trees

Just as branch-and-bound enumeration is visualized by a decision tree (OR tree), AND/OR enumeration can be represented by an AND/OR tree. The cornerstones for AND/OR enumeration in a combinational network as presented in the previous chapter, are the basic notions of recursive learning. Unjustified gates as given by Definition 3.1 and implied signal values represent OR nodes in the AND/OR tree. Justifications as defined in Definition 3.3 represent the AND nodes in the AND/OR tree. In general terms, AND nodes represent some requirements that are either given by the initial set of value assignments or correspond to injected justifications. OR nodes are the logical consequences that result from these requirements. Before formally defining AND/OR reasoning trees, we first discuss AND/OR enumeration which underlies these graphs.

The basic AND/OR enumeration routine is given in Table 4.1. It is derived from routine *make_all_implications()* of Section 3.2.1 after removing the statements to extract implications, i.e., Table 4.1 corresponds to the gray-shaded regions of Table 3.3. The AND/OR enumeration of Table 4.1 solves a satisfiability problem: if the set of value assignments $S$ is logically inconsistent in the circuit (possibly with pre-set value assignments at some nodes), the routine identifies a conflict. If the value assignments are satisfiable that is indicated by the absence of a conflict after exhausting the complete AND/OR tree. This is stated in Theorem 3.1.

Note an important difference from conventional techniques: unlike the common techniques in CAD and testing, this method proves the satisfiability of a set of value assignments in a combinational circuit without actually generating a satisfying input vector. This may only happen in special cases. Similarly, when applied to stuck-at fault test generation, the testability of a fault can be proved without actually generating a test vector. When conventional methods check satisfiability or testability, as a side result, they produce *sufficient solutions*, i.e., inputs that satisfy the function or represent a test vector. In contrast, the AND/OR enumeration based approach described here, as a side result, can generate the *necessary conditions* for a solution in terms of *implications* or *implicants* (see Section 4.3). This reflects the very different nature of the conventional variable enumeration techniques employed in computer-aided circuit design and the AND/OR enumeration developed in this chapter [StKu97].

```
/* this procedure operates on a global data structure representing the
   gate netlist of the circuit with possibly pre-set value assignments
   at some of the nodes, S is a new set of value assignments in the
   circuit, r is the current recursion level, initially r = 0, rmax is a
   user defined aborting criterion  */

and_or_enumerate(S, r, rmax)
{
  /* determine OR nodes of AND/OR tree */
  make all direct implications for S in circuit and
  set up a list U^r of unjustified gates in event list E(S);

  if (value assignments are logically inconsistent)
      return INCONSISTENT;

  /* determine AND nodes of AND/OR tree */
  if (r < rmax)
  {
      for (each unjustified gate g in U^r)
      {
          /* try justifications */
          set up list of justifications gC^r;
          for (each justification Ji ∈ gC^r)
              consistenti := and_or_enumerate(Ji, r+1, rmax );

          /* check logic consistency */
          if (consistenti = INCONSISTENT for all i)
              return INCONSISTENT;
      }
  }
  return CONSISTENT;
}
```

**Table 4.1:** Pseudo-code for AND/OR enumeration in combinational circuit

The difference between conventional variable enumeration and AND/OR enumeration may also be illuminated by the following consideration: conventional variable enumeration is based on making decisions on variable assignments and enumerating all combinations of such assignments (explicitly or implicitly). The variable space in the Boolean domain is always finite. (Otherwise variable enumeration is useless because it would never finish in the case where no solution exists). Suppose that a variable can only assume the two logic values 0 or 1. If the decision for the assignment 1 at some variable has turned out to be wrong then conventional variable enumeration will assign 0. AND/OR enumeration however will not. Note that it is a subtle point which makes it necessary to give special consideration to issues of completeness (see, e.g., the proof of Theorem 3.1). If a certain objective cannot be fulfilled by making this variable 1 then the next enumeration step will assume that it may be fulfilled if some *other* variable is assigned to some value. Rather than exploiting the finiteness of the variable space, AND/OR reasoning exploits the situation that certain conditions must be fulfilled due to the given *structure* of the circuit implementation we are dealing with.

In order to further study the role of AND/OR enumeration in combinational circuits we now introduce AND/OR trees constructed by the routine in Table 4.1. An AND/OR tree is a bipartite tree, one type of node is referred to as AND node, the other type is the OR node. The justifications as performed by *and_or_enumerate()* form the AND-nodes. The situation of value assignments being implied from the justifications represent OR nodes. *Justified* gates are OR nodes without successors, i.e., they are the leaves of the tree. Unjustified gates require justifications and have AND nodes as successors.

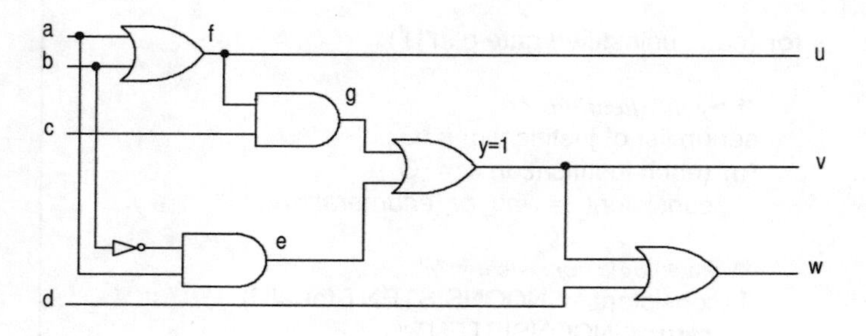


**Figure 4.4:** Example of a combinational circuit with value assignment $y = 1$

**Example 4.3:** Consider the circuit in Figure 4.4. We apply *and_or_-enumerate()* for an initial situation of value assignments $S = \{y = 1\}$. The initial event list is $E = \{y = 1\}$. Node $y$ in the circuit of Figure 4.4 becomes an unjustified line and the complete set of justifications is ${}^{y}C = \{\{g = 1\}, \{e = 1\}\}$. This produces the two AND nodes in level 1 of the AND/OR tree of Figure 4.5. To distinguish AND nodes from OR nodes, AND nodes are marked by an arc. (This is a convention adopted from standard literature.) For each justification direct implications imply logic signal values and produce new unjustified gates. Every value assignment forms an OR node in the tree. For $g = 1$ we imply $c = 1, f = 1$ and $u = 1$, where node $f$ becomes a new unjustified gate. This requires new justifications and the technique continues to enumerate the AND/OR tree as shown in Figure 4.5 in a depth-first manner.

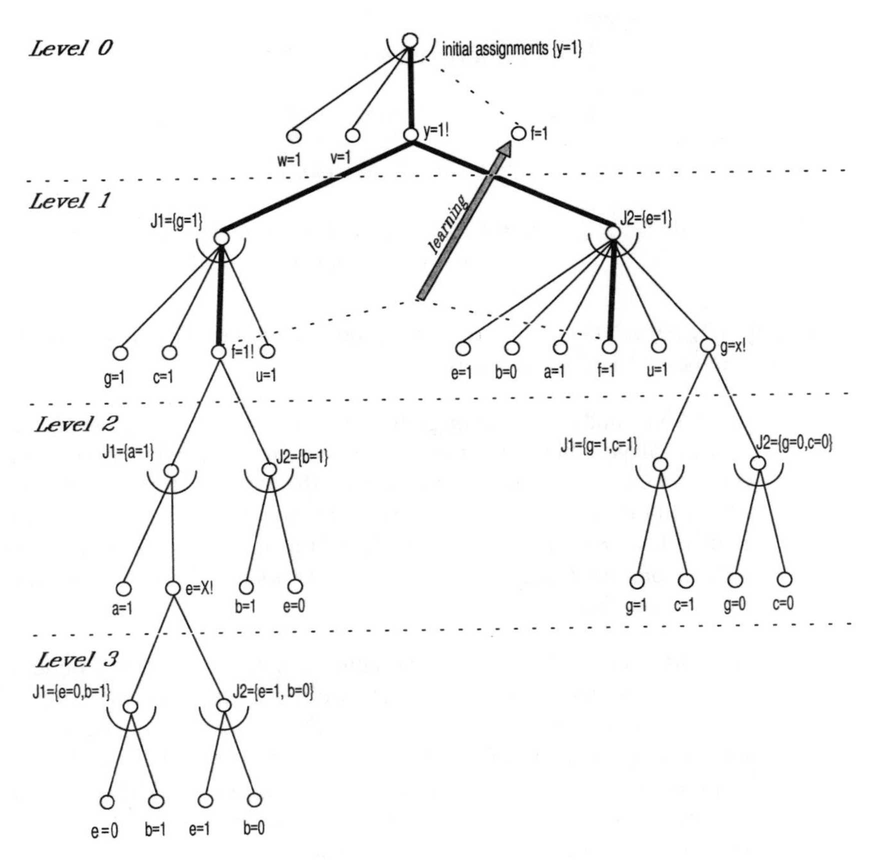


**Figure 4.5:** AND/OR tree for assignment $y = 1$ in the circuit of Figure 4.4

The reader may also examine the examples in Section 3.2 for an illustration of AND/OR enumeration. AND/OR trees are described more precisely by the following definitions:

**Definition 4.1:** An *AND/OR tree* is a bipartite rooted directed tree with two disjoint vertex sets $V_{AND}$ and $V_{OR}$. The root node $v_r$ is an element of $V_{AND}$ . The terminal nodes (*leaves*) of the tree are elements of $V_{OR}$. Adjacent nodes belong to different vertex sets. Each node $v_{OR} \in V_{OR}$ has as attribute a variable assignment $f = V$ where $f$ is an element of a set of variables $\{f_1, f_2, ..., f_n\}$ and $V$ is an element of a set of values $B$. Each node $v_{AND} \in V_{AND}$ has as attribute a set of variable assignments $S = \{f_1 = V_1,\ f_2 = V_2, ..., f_k = V_k\}$. Furthermore, each vertex $v$ has as attribute an integer $level(v)$, such that

i) The root (AND) node $v_r$ has

$$level(v_r) = 0.$$

ii) OR nodes $v_{OR}$ have the same $level(v_{OR})$ as their immediate (AND) predecessors $v_{pred}$:

$$level(v_{OR}) = level(v_{pred}).$$

iii) AND nodes $v_{AND}$ with their immediate (OR) predecessors $v_{pred}$ have

$$level(v_{AND}) = level(v_{pred}) + 1.$$

**Definition 4.2:** An AND/OR tree with root node $v_r$ can be associated with the AND/OR enumeration of Table 4.1 as follows:

i) each AND node $v_{AND}$ belongs to a set $S = \{f_1 = V_1,\ f_2 = V_2,\ \ldots,\ f_k = V_k\}$ of variable assignments at nodes in the combinational network, where this set is given either by the initial set of variable assignments if $v_{AND} = v_r$ (root node), or by justifications for unjustified gates if $v_{AND} \neq v_r$ (intermediate nodes). If a set $S$ turns out to be logically inconsistent, the corresponding AND node and all its successors are removed from the tree.

ii) each OR node $v_{OR}$ belongs to a variable assignment $f = V$ at a node in the combinational network which is required for the logic consistency of the set $S$ associated with the parent AND node of $v_{OR}$, i.e., an OR node belongs to a variable assignment in the event list $E(S)$. If $f = V$ is at the output of an unjustified gate $g$ then $v_{OR}$ has $m$ AND children, each belonging to a justification $J \in {}^gC$, with $m = |{}^gC|$. If $f = V$ is at the output of a justified gate then $v_{OR}$ is a leaf of the tree.

Such a tree is called the *AND/OR reasoning tree* for the initial set of value assignments $S$ and the given combinational network.

For reasons of simplicity we will often speak of AND/OR trees instead of AND/OR reasoning trees and AND/OR graphs instead of AND/OR reasoning graphs. In Section 3.2 our goal was to extract implications. That can be accomplished by checking whether all AND nodes that succeed a given OR node, say $y$, have succeeding OR nodes that all correspond to the same value assignment. If this is the case, these OR nodes with identical value assignments can be attached as OR nodes to the predecessor of $y$. As an example, in Figure 4.5 the two AND nodes corresponding to justifications $\{g = 1\}$ and $\{e = 1\}$ have a succeeding OR node corresponding to $f = 1$. Therefore this OR node can be attached to the predecessor of $y$. This is schematically shown in Figure 4.5. Such shifting of OR nodes has been referred to as "learning" (see Sections 2.7 and 3.2). It can occur in any recursion level and the value assignments resulting in the previous level can change the course of subsequent enumeration so that more logical consequences can be examined faster. The reader may observe this effect at the example of Section 3.1.

The following three aspects are essential to understand the role of AND/OR reasoning graphs and AND/OR enumeration for solving various problems:

i) The main potential of AND/OR graphs comes from the fact that important information about the given problem can be derived without visiting the complete graph. In the above example the implication $y = 1 \Rightarrow f = 1$ can be derived by recursive learning already in the first recursion. Partial graphs can be visited by restricting the maximum recursion depth for *and_or_-enumerate()* to some value $r_{max}$.

ii) In order to evaluate AND/OR graphs with respect to certain information we can traverse them by enumeration without actually building them. Therefore, memory requirements, e.g. for recursive learning, grow linearly with the size of the circuit. However, it is also possible to actually construct the graph. In doing so, trade-offs between memory and time can be achieved. This is briefly discussed in Section 4.4.

iii) The primary goal of AND/OR enumeration in CAD is *not* always to prove or disprove satisfiability. Often, the goal is to extract valuable information from the -possibly not fully completed - enumeration process. This information typically captures the logical consequences of a given set of value assignments or a single stuck-at fault assumption. As will be shown in Chapter 5, topological properties of AND/OR reasoning trees can provide important information for guiding logic optimization.

Recursive learning and AND/OR enumeration have been formulated mainly to be applied in multi-level circuits. Nevertheless, it is illuminating to apply *and_or_-enumerate()* to two-level circuits. Consider the two-level circuit in Figure 4.6. Figure 4.7 shows the AND/OR graph when the value 0 is assigned at the output $y$. AND/OR enumeration for the value 0 at the output of a two-level SOP-type circuit performs a *tautology test*. The SOP is a tautology if and only if a conflict is produced by *and_or_enumerate()*. As can be noted, the AND/OR tree for a *unate* SOP is very simple and has the same structure as the two-level circuit. The root AND node in the AND/OR tree corresponds to the OR gate in the circuit and the succeeding OR nodes correspond to the AND gates in the circuit. Obviously, this is because the AND gates in the circuit represent implicants for function $y$ and therefore the value assignment $y = 0$ implies OR nodes ($h = 0$, $i = 0$, $j = 0$) which correspond to these implicants. Since the circuit implements a unate function the AND/OR tree terminates in the next level. All AND nodes have only one succeeding OR node, representing a leaf of the tree. This reflects the well-known fact that tautology checking in unate functions is of polynomial complexity. Note that it is only in this special case that the AND nodes and OR nodes of the AND/OR tree have direct relationships with the OR *gates* and AND *gates* of the circuit.

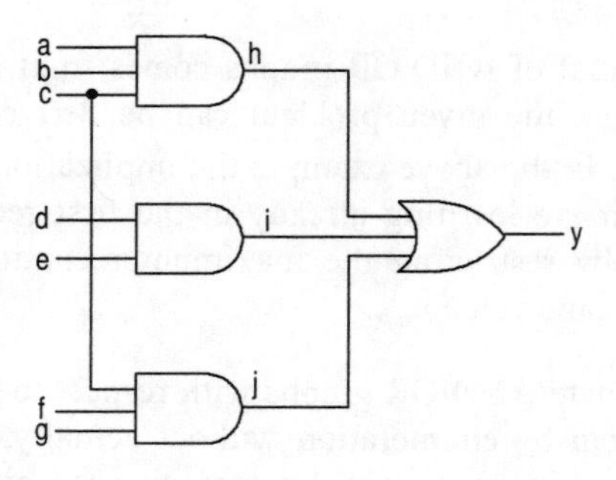


**Figure 4.6:** A two-level circuit for a unate function

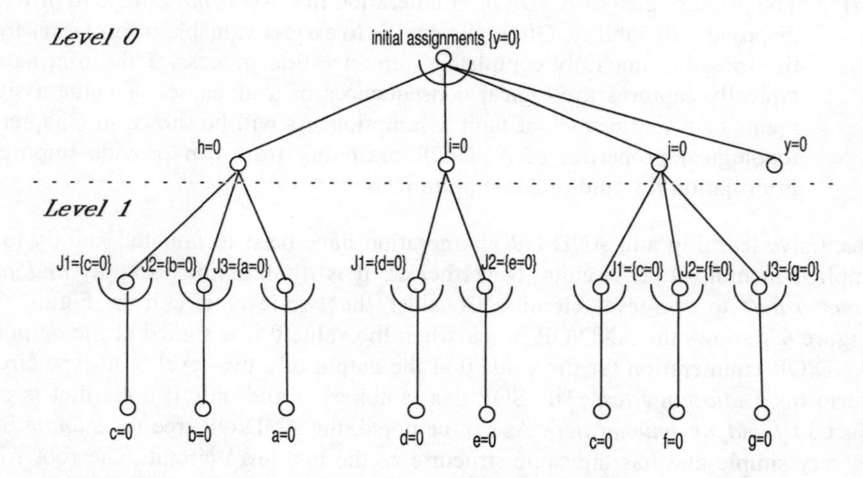


**Figure 4.7:** AND/OR tree for the unate circuit of Figure 4.6

Let the AND/OR tree be levelized according to the recursion depth *r* in *and_or_enumerate()*, then each level consists of a set of AND nodes with their OR successors. The following theorem holds:

**Theorem 4.1:** Let $y$ be the output signal of a two-level combinational circuit in SOP-form. The AND/OR tree for the assignment $y = 0$ (tautology test) has only 2 levels if the SOP-expression is unate (see Appendix for a proof).

The fact that the AND/OR tree for a unate SOP has only two levels is also related to the well-known result that all prime implicants in a unate SOP are essential, i.e., the unate SOP is a syllogistic formula [Brow90]. If the circuit is not unate the AND/OR tree has to be continued after level 1 in order to explore the logic consequences which are not covered by the implicants being included in the SOP.

The situation for the non-unate case is illustrated in Figure 4.8 and Figure 4.9 where the circuit of Figure 4.6 is modified such that it becomes non-unate in variable $c$.

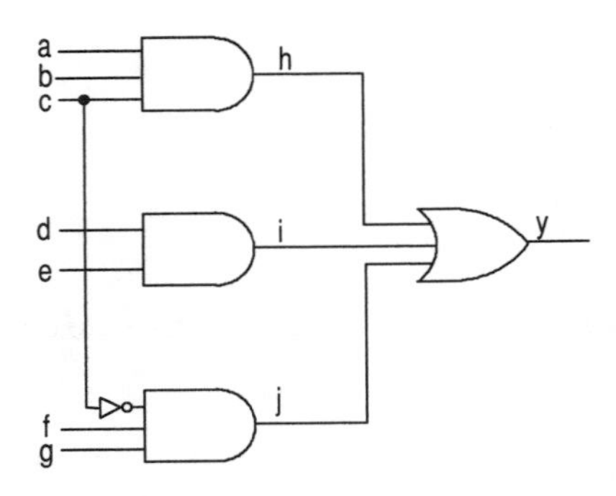


**Figure 4.8:** Non-unate circuit

Also in the case of a non-unate circuit, level 0 of the AND/OR tree reflects the implicants in the SOP. If the circuit is not unate however, the AND/OR tree continues after level 1. This is because the justifications at some unjustified line, e.g., $h = 0$ in Figure 4.8, produce events at other unjustified lines without justifying them. The justification $c = 0$ at gate $h$ produces a logic 1 at the input of gate $j$. This changes the status at gate $j$ and represents an event so that the unjustified line $j = 0$ is added to the list of unjustified gates for the next recursion level. As can be noted, destroying the unateness of variable $c$ by adding an inverter as shown in Figure 4.8 leads to an AND/OR tree with three levels as shown in Figure 4.9.

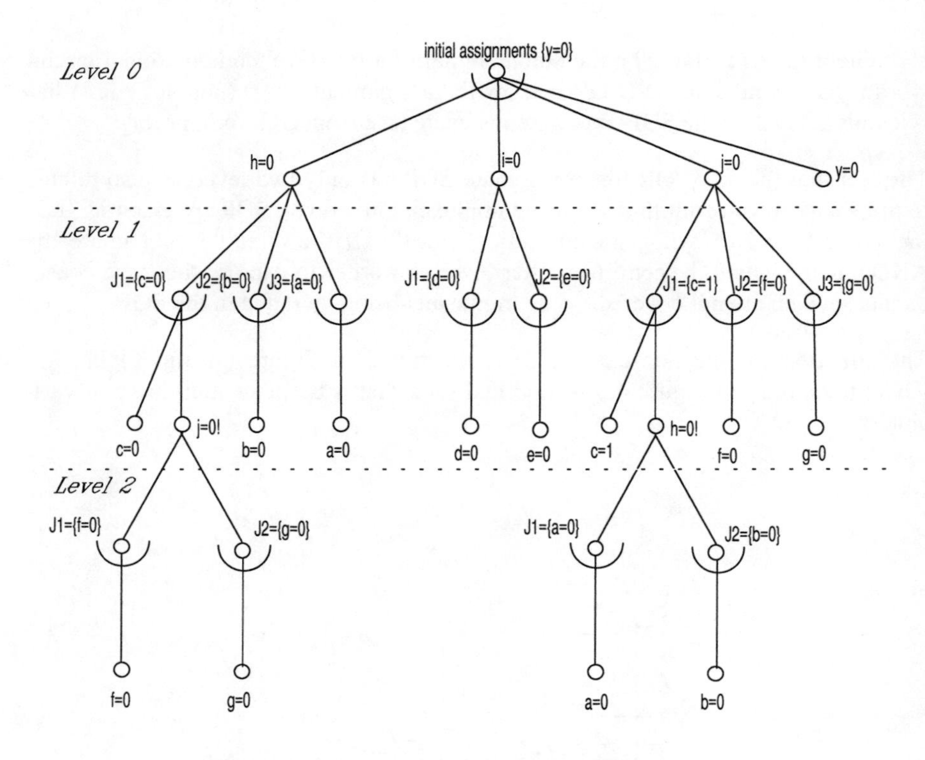


**Figure 4.9:** AND/OR tree for non-unate circuit

## 4.3 Implicants in Multi-Level Circuits

The notions of *implicants* and *prime implicants* of Boolean functions are well-known from two-level minimization theory. Many methods have been developed in the past to calculate the prime implicants of a Boolean function [McCl86]. However, when dealing with general, multi-level combinational circuits the basic concepts of implicants and prime implicants are rarely used. Early attempts to develop multi-level optimization procedures based on extensions to the Quine-McCluskey procedure [Lawl64] proved impractical for circuits of realistic size. The classical two-level minimization concepts can be applied to internal nodes or local regions of the network [Bray87], but this alone is not sufficient and a variety of additional techniques is needed to take into account the structural properties of the given network. Therefore, in this section, we extend the meaning of the notion of prime implicants for broader use in multi-level circuits. As already illustrated in Example 4.1 we express implicants in multi-level circuits not only in terms of the variables of the function at the considered node but in terms of arbitrary nodes in the network, including those which are not in the transitive fanin of the considered node function.

Actually, methods to identify such generalized implicants in multi-level networks have already been developed in the field of automatic test pattern generation. The implication procedures described in Sections 2.3, 2.7 and 3.2.1 can be interpreted as algorithms to calculate (prime) implicants. Note however that the limitation of such implication techniques is that they only generate implicants that can be expressed in terms of a *single* literal belonging to a specific node in the network. Based on AND/OR reasoning trees we now describe a method to identify *multi-literal* implicants. The focus of this section is on relating certain subtrees of the AND/OR reasoning trees to implicants and prime implicants in the network. An application to multi-level circuit minimization will be considered in Section 5.6.

### 4.3.1 Prime Implicants

Several notions from two-level logic minimization are now extended to describe our methods for multi-level logic circuits. A *literal* is a variable in the *combinational network* or its complement. A *product term* is a conjunction of literals. For example $a$, $\bar{a}$, $b$, $\bar{b}$, $c$, $\bar{c}$.... are literals, $abc$, $a\bar{b}c$, $\bar{a}\bar{b}c$ ... are product terms.

In two-level minimization circuits are usually represented as sum of products (SOP), see Section 1.3.2, and the notions and algorithms of two-level minimization are typically tailored for such SOP expressions. Of course, there always exists a dual representation of the switching function in a product of sums (POS) form with a dual set of notions and algorithms. Note that in the case of a general multi-level network it is not sufficient to restrict notions and algorithms to only one of the two dual representations because both are equally important and both have to be considered in order to achieve an optimum implementation. For this reason, the term *implicant* as used in two-level minimization has to be refined as follows:

**Definition 4.3:** A *1-implicant* (*0-implicant*) for a given function $y$ in a combinational network $C$ is a product term $t$ such that $y$ assumes the value 1 (0) for every set of value assignments at the primary inputs of $C$ for which $t$ assumes the value 1. An *implicant* for a node $y$ is product term which is either a 0-implicant or a 1-implicant for $y$.

Note that 1-implicants correspond to the classical notion of implicants as used in the theory of two-level minimization.

**Definition 4.4:** An *implicant* is called *prime* if the removal of any literal makes the implicant a product term that is not an implicant.

As already explained, the literals of an implicant for some node $y$ in a *multi-level* combinational network can belong to arbitrary nodes in the network. We will now examine how such implicants in combinational networks can be derived by AND/OR reasoning trees. It has been shown in Section 3.2.1 how all *single-literal* implicants can be extracted from the AND/OR enumeration process (= recursive learning). This is now extended to extract arbitrary, *multi-literal* implicants.

The following definition relates implications and implicants in a combinational network to certain subtrees of the AND/OR enumeration tree defined in Section 4.2.

**Definition 4.5:** An *implication subtree* (*IST*) is an AND/OR tree with the following properties:

i) it is a subtree of an AND/OR reasoning tree,
ii) the enumeration tree and its subtree have the same root node,
iii) for each AND node included in the subtree, all its siblings in the AND/OR reasoning tree are also included in the subtree.

As an example, the bold lines in Figure 4.5 indicate an IST.

**Theorem 4.2:** Let $y$ be an arbitrary node in a combinational network and $T$ be the AND/OR enumeration tree for an initial set of value assignments $S = \{y = 0\}$. Consider a product term $t = x_1 \cdot x_2 \cdot ...x_m$ where $x_i$ is a literal corresponding to a variable $f_i$ or its complement in the combinational network. Further, consider an IST of $T$ with a set of leaves $L$.

If there is a one-to-one mapping between the literals $x_i$ of $t$ and the elements $(f_i = V_i)$ of $L$ such that $V_i = 0$ if $x_i$ represents the uncomplemented variable $f_i$, and $V_i = 1$ if $x_i$ represents the complemented variable $\overline{f_i}$, then $t$ is a 1-implicant for $y$. Analogously, $t$ is a 0-implicant for $y$ if the IST is a subtree of the enumeration tree with the initial assignment $S = \{y = 1\}$ (see Appendix for a proof).

Theorem 4.2 states the rule for deriving implicants from an AND/OR tree. An implicant is formed by the conjunction of variables belonging to the leaves of an IST.

If a variable at a leaf of the IST is assigned to 0 then we have to take the uncomplemented variable, if it is assigned to 1 we have to take the complemented variable as a literal in the implicant.

Figure 4.5 shows an example for a single-literal implicant. As mentioned above, implications as performed in test generation can be related to single-literal implicants. If the basic implication routine in the AND/OR enumeration process performs direct implications, then, in the AND/OR tree, direct implications correspond to an IST completely contained in level 0, i.e., the IST only consists of the root node with one or several of its immediate OR successors. For an *indirect* implication, say $y = 1 \Rightarrow f = 1$, there must exist an IST with root node $y$ and an initial set of value assignments $\{y = 1\}$, which also includes OR nodes at deeper levels of the AND/OR tree and fulfills the condition that *all* leaves belong to the same value assignment $f = 1$. The bold lines in Figure 4.5 indicate an IST, which corresponds to the indirect implication $y = 1 \Rightarrow f = 1$. Using the notions of Definition 4.3 and applying Theorem 4.2 this can also be stated by saying that $\overline{f}$ is a (single-literal) 0-implicant for $y$ in the circuit of Figure 4.4. An example of multi-literal implicants will follow in Section 4.3.2.

How are *prime* implicants represented in the AND/OR tree? Note in Definition 4.5 there is no requirement to include in the IST more than *one* child of each AND node of the original tree. In fact, including more than one OR child of any AND node makes the IST non-minimal. Since this non-minimal IST can contain leaves with new variable assignments not needed to make the product term an implicant, the corresponding implicant is non-prime.

**Definition 4.6:** An IST is called *minimal implication subtree* (*MIST*) if each AND node has exactly one OR child.

**Theorem 4.3:** Let $y$ be an arbitrary node in a combinational network and $T$ be the AND/OR reasoning tree for an initial set of value assignments $S = \{y = V\}$, $V \in \{0, 1\}$. For every *prime* implicant of $y$ there exists a minimal implication subtree (MIST) of $T$ such that the leaves of the MIST correspond to the literals of the prime implicant as given in Theorem 4.2 (see Appendix for a proof).

Note that not every MIST corresponds to a prime implicant. For a given MIST with a set of leaves $L$ there may be some other MIST with a set of leaves $L'$ such that $L' \subset L$. Obviously, then, the implicant belonging to the first MIST cannot be prime. Fortunately, by tracing from the leaves towards the root of the AND/OR tree, it is very easy to check for a given MIST with a set of leaves, whether a subset of these leaves can belong to another MIST.

### 4.3.2 Permissible Prime Implicants

Often, the value of an internal logic function cannot be *observed* at any of the circuit outputs for certain sets of input assignments and, therefore, in such situations the value of the function need not be specified. This leads to so called *observability don't cares* (see Section 5.3.4). Any function that covers such an incompletely specified function is called a *permissible function* according to Muroga [Muro89].

> **Definition 4.7:** In a combinational network a function $y'$ is called *permissible* at a node with function $y$, if the function $C(\boldsymbol{x})$ of the combinational network does not change when $y$ is replaced by $y'$.

Since we extend the notion of implicants from two-level circuits to multi-level circuits it seems wise to also take into account the concepts of observability don't cares and permissible functions. Remember that making implications or determining implicants means to calculate necessary conditions for some function to be satisfiable. In the two-level domain, the considered functions always belong to the primary outputs of the circuit. Observability is therefore not an issue. For multi-level circuits, however, we ask for the satisfiability or *controllability* of some *internal* function of the network. If we also include *observability* into our consideration this leads to a further extension of the notion of an implicant that turns out to be very useful in practice:

> **Definition 4.8:** For some node $y$ in a combinational network $C$, a product term $t$ of some node variables of $C$ is called a *permissible 1-implicant* for $y$, if and only if the following condition holds: If $t$ is 1 then $y$ is 1 or $y$ is *not observable* at any primary output of $C$. Similarly, $t$ is called a *permissible 0-implicant* for $y$, if and only if the following condition holds: If $t$ is 1 then $y$ is 0 or $y$ is *not observable* at any primary output of $C$. A permissible implicant is called *prime* if the removal of any literal makes the permissible implicant a product term that is not a permissible implicant.

AND/OR enumeration for a given node in the combinational network can take into account observability don't cares. Besides considering justifications at unjustified gates as described in Section 4.2, sensitizations at the D-frontier can also be included in the enumeration process for a given stuck-at fault assumption. This involves using Roth's D-alphabet $B_5 = (0, 1, X, D, \overline{D})$ instead of the conventional three-valued system $B_3 = (0, 1, X)$. The AND/OR tree associated with these extensions starts with a single stuck-at fault assumption and, in the sequel, is referred to as *D-AND/OR reasoning tree*. Analogously, as the AND/OR enumeration of Table 4.1 results from *make_all_implications()* by removing the statements to extract necessary assignments, D-AND/OR enumeration results from *com-*

*plete_unique_sensitization()* in combination with *make_all_implications()* of Section 3.2.2.

**Theorem 4.4:** Let $y$ be an arbitrary node in a combinational network and $T$ be the D-AND/OR enumeration tree for a fault, $y$ stuck-at-1. Consider a product term $t = x_1 \cdot x_2 \cdot \ldots x_m$ where $x_i$ is a literal corresponding to a variable $f_i$ or its complement in the combinational network. Further, consider an IST of $T$ with a set of leaves $L$ such that in the combinational network the nodes $f_i$ *cannot be reached by the fault effect.*

If there is a one-to-one mapping between the literals $x_i$ of $t$ and the elements $(f_i = V_i)$ of $L$ such that $V_i = 0$ if $x_i$ represents the uncomplemented variable $f_i$, and $V_i = 1$ if $x_i$ represents the complemented variable $\overline{f_i}$, then $t$ is a permissible 1-implicant for $y$. Analogously, $t$ is a permissible 0-implicant for $y$ if the IST is a subtree of the enumeration tree for $y$ stuck-at-0 (see Appendix for a proof).

The advantage of D-AND/OR enumeration over the AND/OR enumeration of Table 4.1 is that it incorporates *permissible functions* into the reasoning. Theorem 4.5 states an important property of AND/OR trees which makes them attractive in logic synthesis:

**Theorem 4.5:** Let $y$ be an arbitrary node in a combinational network and $T$ be the D-AND/OR enumeration tree for a fault, $y$ stuck-at-$V$, $V \in \{0, 1\}$. For *every permissible prime implicant* at node $y$ there exists a minimal implication subtree (MIST) of $T$ such that the leaves of the MIST correspond to the literals of the prime implicant as given in Theorem 4.4 (see Appendix for a proof).

**Example 4.4:** Figure 4.10 shows a circuit for which the D-AND/OR reasoning tree is built in Figure 4.11. Consider the fault $a$ stuck-at-1. There are two paths along which this fault can propagate to a primary output. At least one of them has to be sensitized for fault detection. One path traverses gates $k$ and $l$. Its sensitization yields the value assignments $d = 1$ and $j = 0$. For the AND/OR tree in Figure 4.11, this produces the left AND node in level 1 with its children. The second possibility is to sensitize the path through $m$ and $q$ resulting in the right portion of the AND/OR tree. The sensitizations yield value assignments and unjustified lines. These value assignments are enumerated in the usual way as given by Table 4.1, so that the AND/OR tree for the stuck-at-1 fault at signal $a$ is as shown in Figure 4.11. Note that for reasons of simplicity we only consider unjustified gates with specified output signals, i.e., the gates referred to as unjustified lines in test generation literature, for inclusion in the AND/OR tree. Although unjustified gates with unspecified outputs as in the AND/OR tree of Figure 4.5 are necessary for the theoretical completeness of the enumeration, it is possible to neglect them for most practical purposes [KuPr94]. The bold lines in Figure 4.11 indicate a MIST that represents a permissible prime implicant $b \cdot c$ for node $a$ in the circuit. We will return to this example later in Section 5.6.2 .

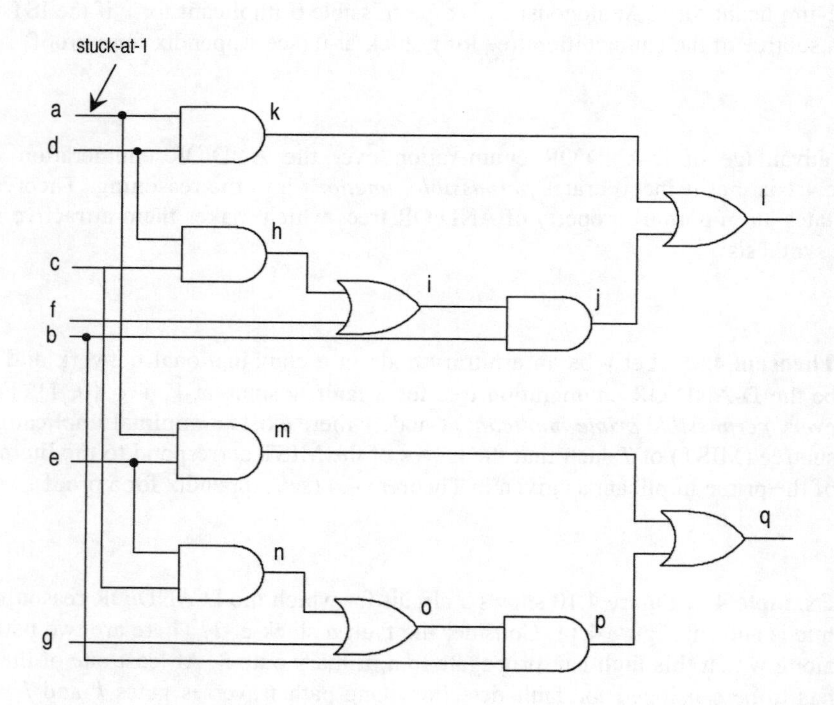


**Figure 4.10:** Example circuit for optimization

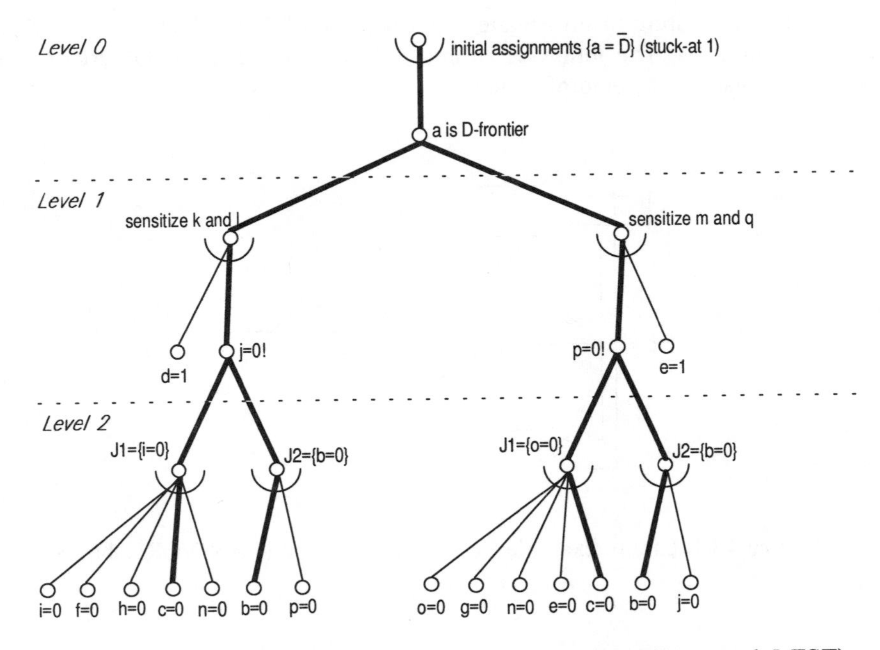


**Figure 4.11:** AND/OR tree for circuit in Figure 4.10 (bold lines mark MIST)

## 4.4 Search State Hashing and Isomorphic Subtrees

A representation of Boolean functions by AND/OR reasoning graphs is of a totally different nature when compared to binary decision diagrams. It is a non-canonical representation of a Boolean function or a combinational circuit, i.e., the size and shape of the graph depend on the structure of the circuit from which it is built. While canonicity is an advantage in some applications like formal verification there are other applications where it is of advantage to maintain some structural information about the given implementation. Such applications are particularly in the domain of logic synthesis.

So far we have investigated AND/OR enumeration as an alternative to the classical variable enumeration techniques. If the classical variable enumeration is represented by a Shannon tree significant reduction can be achieved for a wide variety of circuits by sharing isomorphic subtrees so that a binary decision diagram is obtained. It is interesting to investigate whether such a "BDD-effect" can also occur for AND/OR trees. The following example shows how an AND/OR tree can be reduced by sharing of isomorphic subtrees.

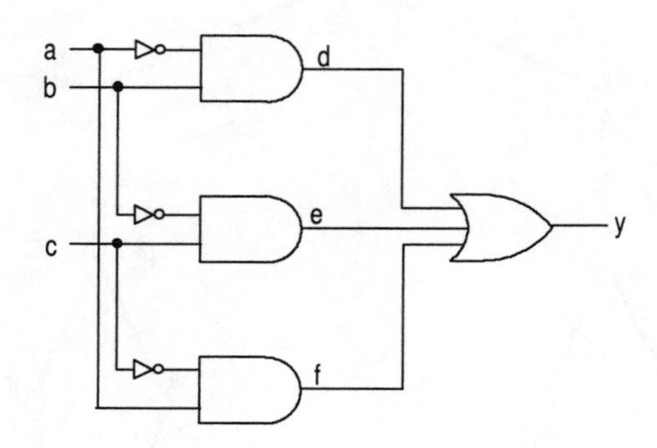


**Figure 4.12:** Circuit example for sharing isomorphic AND/OR subtrees

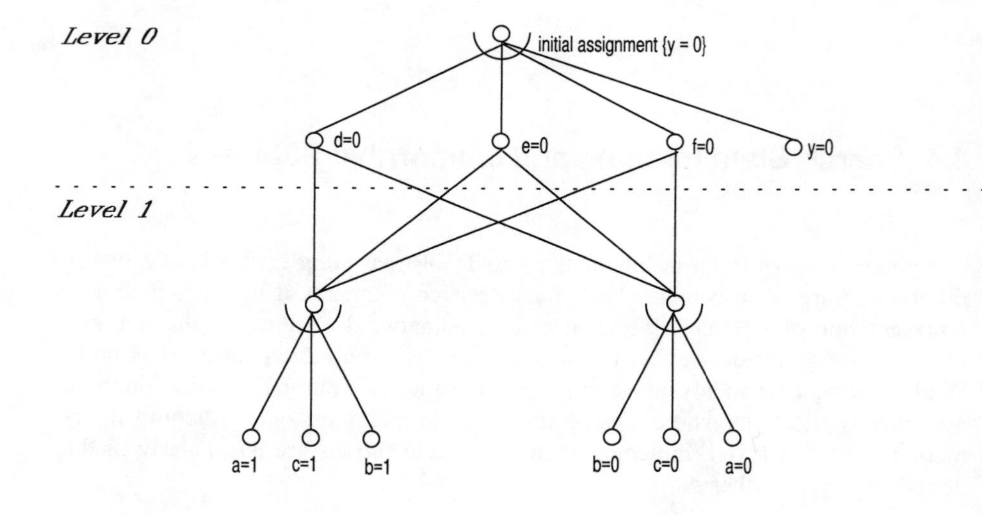


**Figure 4.13:** AND/OR graph for circuit in Figure 4.12

Consider the circuit in Figure 4.12. The AND/OR tree for $y = 0$ can be constructed using the AND/OR enumeration of Table 4.1. If isomorphic subtrees are shared we obtain the AND/OR graph of Figure 4.13.

In general it is desirable to share isomorphic subtrees. So far we have only extracted information from AND/OR trees by enumerating them. If the AND/OR trees are actually built and reduced as shown in Figure 4.13, a significant amount of enumeration effort may be saved. Similarly, just as OBDDs provide substantially more power than exhaustive simulation it can be expected that appropriate trade-offs between time and memory (using hashing and caching techniques) can further improve the performance of AND/OR graph based methods.

Besides hashing for isomorphic subtrees there exist many other possibilities to improve the efficiency of the AND/OR search. Similar concepts as have been developed by [ChBu96] for branch-and-bound methods can also be used to improve the efficiency of AND/OR search. Future work may investigate this with particular focus on finding appropriate trade-offs between time and memory.

# Chapter 5

# LOGIC OPTIMIZATION

Multi-level logic optimization figures prominently in the synthesis of very large integrated circuits. Sections 5.1 to 5.3 briefly review basic approaches in multi-level logic optimization. The main objective of this chapter is to demonstrate the intimate relationship between algorithmic concepts of test generation, the AND/OR reasoning of Chapters 3 and 4 and common notions of logic synthesis. Section 5.6 develops a new approach to multi-level logic optimization based on the methods presented in Chapters 3 and 4.

## 5.1 Problem Formulation

The goal of multi-level logic optimization is to transform an arbitrary combinational circuit $C$ into a functionally equivalent circuit $C'$, circuit $C'$ being less expensive than $C$ according to some cost function. The cost function typically incorporates

- area
- speed
- power consumption
- testability

as the main objectives of the optimization procedure. This chapter presents concepts and algorithms which are mostly independent of the specific cost function and are basic ingredients for a logic synthesis tool, no matter which of the above criteria is considered the most important. The specific formulation of the cost function may affect the heuristic guidance of the optimization tool, or may require techniques in addition to the ones to be presented, but it generally does not affect the validity of the basic algorithms. Whenever *heuristics* are discussed in this chapter, we will assume that *minimum area* is the primary objective of the optimization procedure.

There are many possibilities for estimating the area requirements of a physical circuit implementation if the circuit is described at the gate-level. For a technology independent circuit representation, two different measures are frequently used in the literature: *literal count* and *connection count*. The former is typically used when the circuit is represented by a general *Boolean network* as defined in Section 1.3.3. Some recent approaches, like the ones to be described, use a restricted Boolean network as the basis for all operations. This restricted Boolean network, described in Section 1.3.3 corresponds more directly to a typical gate-level netlist description of the circuit and is referred to as *combinational network* in this book. If the circuit is represented by a combinational network the common area measure is the *connection count*. The difference is explained by the following example.

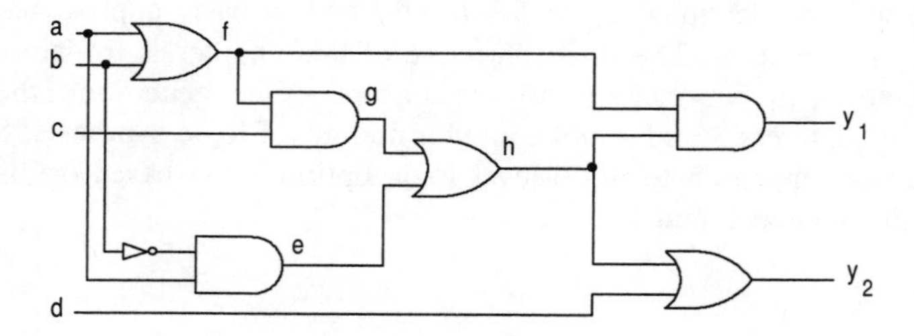


**Figure 5.1:** Example circuit to illustrate literal count versus connection count

**Example 5.1:**

Consider the circuit in Figure 5.1. It can be described by the following set of Boolean expressions:

$$f = a + b$$
$$h = f \cdot c + a \cdot \bar{b}$$
$$y_1 = f \cdot h$$
$$y_2 = h + d$$

This set of Boolean expressions corresponds to a representation of the circuit as a Boolean network. The corresponding Boolean network is shown in Figure 5.2. As defined in Section 1.3 a Boolean expression consists of *literals*. The size of a Boolean network is measured by the number of literals needed to describe each function at individual nodes summed over all nodes in the network. For the representation in Figure 5.2 we count 10 literals (we only count the right side of each equation). Obviously, the literal count depends on how the circuit description is "collapsed" into the nodes of the Boolean network and how the function at each node is represented. For example, if the function at a node is represented in a factored form (see Section 1.3.2) the literal count will usually be smaller than if the node function is represented in the SOP (sum of products) form. Therefore, when evaluating optimization results with respect to literal counts, great care has to be taken in order to ensure a fair comparison.

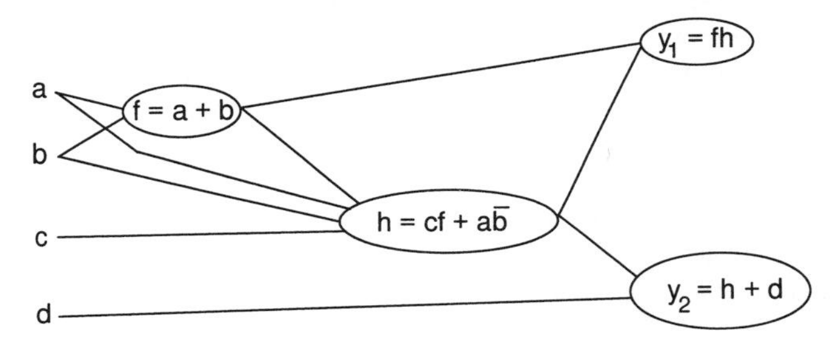


**Figure 5.2:** Boolean network for circuit in Figure 5.1

If the circuit is represented by a combinational network as shown in Figure 5.1, it is common to count connections. A *connection* is defined to be a distinct input of a gate having at least two inputs, i.e., inputs to inverters are not counted. Inverters are not counted because most optimization techniques disregard the costs for inverters during the main optimization procedure and minimize the number of inverters in a post-processing phase. This step is often called *phase assignment*. Furthermore, neglecting inverters when counting connections is analogous to treating a variable and its complement equally as one literal when determining the literal count. In the above example, we count 12 connections.

## 5.2 Functional Decomposition

The earliest systematic approach to multi-level logic minimization is known as *functional decomposition* and was formulated in the 1960s by Ashenhurst

[Ashe59], Curtis[Curt61] and Roth and Karp [RoKa62]. Functional decomposition is the process of re-expressing a single switching function of $n$ variables by a collection of $m$ functions such that $m < n$, and each function depends on less than $n$ variables. Repeating this process for the created sub-functions produces a multi-level representation of the given switching function. This has direct applications to look-up table (LUT) based field programmable gate arrays (FPGAs). There, a given switching function is broken down into a number of simpler sub-functions so that it can be implemented by the basic blocks of the FPGA. With FPGAs becoming more popular there is also a renewed interest in functional decomposition. This section summarizes some results from the classical theory which form the basis of recent applications and recent research activities, e.g., [HwOw90], [LaPe93], [Sasa93], [MuBr94], [StSt94], [MoSc94] and [WuEc95], [LaPa96].

In the following, only completely specified switching functions are considered. A switching function $f$: $B^n \rightarrow B$, with $B = \{0, 1\}$ is said to be *decomposable*, if its $n$ variables can be grouped into two sets corresponding to vectors $\boldsymbol{x}$ and $\boldsymbol{y}$ such that

$$f(\boldsymbol{x}, \boldsymbol{y}) = g(\varphi_1(x_1, x_2,\dots x_s), \varphi_2(x_1, x_2,\dots x_s),\dots \varphi_m(x_1, x_2,\dots x_s), y_1, y_2,\dots y_{n-s}) \quad \text{(Eq. 5.1)}$$

If the vectors $\boldsymbol{x}$ and $\boldsymbol{y}$ do not contain any common variables the decomposition is called *disjoint*. The functions $\varphi_i$ are called decomposition functions. The decomposition is called *simple* if the function is decomposed using a single decomposition function $\varphi$, i.e., $m = 1$. The set of variables $\{x_1, x_2,\dots x_s\}$ is called the *bound set* (BS) and the variable set $\{y_1, y_2,\dots y_{n-s}\}$ is called the *free set* (FS).

Traditionally, functional decomposition is accomplished using *decomposition charts.* A decomposition chart for a switching function is similar to a Karnaugh map but the rows and columns are labeled in a different way. In a decomposition chart the columns are labeled in decimal order by the $2^{n-s}$ combinations of binary value assignments to the variables of the bound set. The rows are labeled in decimal order by the variables of the free set. Each entry corresponds to a unique minterm of the function and denotes whether the minterm belongs to the on-set or the off-set of the function. Decomposition chart and the basic idea of functional decomposition are illustrated in the following example.

**Example 5.2:** Take a function, $f = \bar{a}\bar{b}\bar{c} + ab\bar{c} + \bar{a}bcd + a\bar{b}cd$. A simple decomposition is to be determined. We choose BS = $\{a, b\}$ and FS = $\{c, d\}$, i.e., we look for a decomposition of the form $f(a, b, c, d) = g(\varphi(a, b), c, d)$. The decomposition chart is shown in Figure 5.3.

Each cell in the matrix denotes a unique minterm of $f$, i.e., each minterm is determined by a specific row and specific column. Note that there are only two *distinct* columns in the chart of Figure 5.3.

| $\frac{ab}{cd}$ | 00 | 01 | 10 | 11 |
|---|---|---|---|---|
| 00 | 1 | 0 | 0 | 1 |
| 01 | 1 | 0 | 0 | 1 |
| 10 | 0 | 0 | 0 | 0 |
| 11 | 0 | 1 | 1 | 0 |

**Figure 5.3:** Decomposition chart for $f$

Two columns in the decomposition chart are said to be equivalent if they contain the same patterns of zeros and ones. This defines an *equivalence relation* $R_f$ on the columns of the decomposition chart. The equivalence classes consist of the columns with identical patterns. In this section only completely specified functions are considered. For incompletely specified functions these notions must be extended to *compatibility relations*.

**Example 5.2 (contd.):** For the decomposition chart in Figure 5.3 the equivalence classes are $C_1 = \{00, 11\}$ and $C_2 = \{01, 10\}$. Since the equivalence classes belong to unique column patterns a minterm of function $f$ is uniquely determined by the specific row in the decomposition chart and by the equivalence class containing the respective column. This observation is the key to decomposing $f$ in the desired way. Since there are only two equivalence classes it is possible to define a function $\varphi$ in terms of the variables of the bound set and to associate one equivalence class with its on-set and the other equivalence class with its off-set. Let $C_1$ be associated with the on-set, i.e., $\varphi(a, b) = \bar{a}\bar{b} + ab$. Function $f$ can now be described by collecting all on-set minterms of $f$ from the decomposition chart and expressing them in terms of $\varphi$ or $\bar{\varphi}$ (specifying column) and a combination of variables $c$ and $d$ (specifying row). In this example, $f(\varphi, c, d) = \varphi\bar{c}\bar{d} + \varphi\bar{c}d + \bar{\varphi}cd$ and a simple disjoint decomposition of $f$ is obtained in terms of a decomposition function $\varphi$. Further simplification of this expression yields $f(\varphi, c, d) = \varphi\bar{c} + \bar{\varphi}cd$. Note that this decomposition is only possible because all columns of the decomposition chart are covered by no more than two equivalence classes which can be associated with function $\varphi$ and its complement.

The number of distinct columns is called *column multiplicity*, $v$. The observations in the above example are summarized in the following decomposability criterion:

**Theorem 5.1 (Ashenhurst):** A function $f$ is simply and disjointly decomposable iff the column multiplicity $v$ of the decomposition chart for $f$ is at most two, $v \leq 2$.

As illustrated in the example, a simple disjoint decomposition for a function $f$ is accomplished by associating the equivalence classes of columns in the decomposition chart uniquely with the on-set or off-set of the decomposition function $\varphi$. In the symbolic expression for $f$ the variables of the bound set are then replaced by $\varphi$ or $\overline{\varphi}$. Therefore, a decomposition exists only if there are no more than two classes. Note that it does not matter which equivalence class is associated with $\varphi$ and which is associated with $\overline{\varphi}$. In either case, a valid decomposition is obtained. The choice of associating an equivalence class either with $\varphi$ or $\overline{\varphi}$ can also be referred to as an *encoding* of the equivalence classes. Obviously, for the case of simple decomposition the code length is one.

Next, we study the case of a *non-simple* disjoint decomposition, i.e., when there are several decomposition functions. This provides a larger code space and more equivalence classes can be encoded. Obviously, a set of $m$ decomposition functions can be used for the encoding of a maximum of $2^m$ equivalence classes. Therefore, a function $f$ is disjointly decomposable with $m$ decomposition functions iff the column multiplicity $v$ of the decomposition chart for $f$ is at most $2^m$, $v \leq 2^m$. This generalization of Theorem 5.1 was proposed by Roth and Karp [RoKa62].

For non-simple decomposition we can also proceed as in the above example. The decomposition functions are chosen such that each equivalence class of columns in the decomposition chart is assigned to a unique combination of on-sets and off-sets of decomposition functions. This is a natural generalization of the procedure illustrated in Example 5.2. Note however that the equivalence classes do not *have to* be encoded in such a *strict* way [RoKa62]. A looser requirement for choosing codes is given by the following theorem.

**Theorem 5.2:** Let $B^s$ with $B = \{0, 1\}$ be the set of all vertices defined by the $s$ variables of the bound set and let $\boldsymbol{x_1} \in B^s$ and $\boldsymbol{x_2} \in B^s$. Further let $R_f$ be the equivalence relation as above on the columns of the decomposition chart which are labeled by the vertices of $B^s$. Given a function $f$ and an $m$-ary decomposition $\varphi$, function $f$ is disjointly decomposable for $\varphi$ iff

$$\neg(\boldsymbol{x_1} R_f \boldsymbol{x_2}) \Rightarrow \varphi(\boldsymbol{x_1}) \neq \varphi(\boldsymbol{x_2})$$

Clearly, according to Theorem 5.2 the decomposition functions must be chosen such that columns with different patterns will belong to different code words, i.e., they must be assigned to a different combination of on-sets and off-sets of decomposition functions. If the columns have the same patterns they can have the same or a different code word. In the case of a different encoding the decomposition is called *non-strict* [Karp63]. Note an important difference between a strict and a non-strict decomposition. While strict decompositions result from encoding the equivalence classes (or compatibility classes), non-strict decompositions, more generally, can individually encode the minterms. It only has to be assured that the decomposition condition of Theorem 5.2 is fulfilled. A non-simple, non-strict disjoint decomposition is illustrated in the following example.

**Example 5.3:** Consider the switching function $f$ given by the decomposition chart of Figure 5.4. The columns can be partitioned into following equivalence classes, $C_1 = \{000, 001, 100\}$, $C_2 = \{010\}$ and $C_3 = \{011, 101, 110, 111\}$.

| *abc* / *de* | 000 | 001 | 010 | 011 | 100 | 101 | 110 | 111 |
|---|---|---|---|---|---|---|---|---|
| 00 | 1 | 1 | 0 | 0 | 1 | 0 | 0 | 0 |
| 01 | 0 | 0 | 1 | 0 | 0 | 0 | 0 | 0 |
| 10 | 0 | 0 | 0 | 1 | 0 | 1 | 1 | 1 |
| 11 | 1 | 1 | 1 | 1 | 1 | 1 | 1 | 1 |

**Figure 5.4:** Decomposition chart

Since there are three equivalence classes, $v = 3$, and at least two decomposition functions, $\varphi_1$ and $\varphi_2$, are needed for a disjoint decomposition. The three equivalence classes can be assigned to the decomposition functions, for example, $\varphi = (00)$ for $(a\ b\ c) \in \{000, 001, 100\}$, $\varphi = (01)$ for $(a\ b\ c) \in \{010\}$, $\varphi = (10)$ for $(a\ b\ c) \in \{011, 101, 111\}$, and $\varphi = (11)$ for $(a\ b\ c) \in \{110\}$. Note that this encoding fulfills the criterion of Theorem 5.2. With this encoding the columns labeled 011, 101, 111 and 110 are assigned to the on-set of $\varphi_1$ and the columns labeled 010 and 110 are assigned to the on-set of $\varphi_2$. Hence, the resulting decomposition functions are $\varphi_1 = \bar{a}bc + a\bar{b}c + abc + ab\bar{c}$ and $\varphi_2 = \bar{a}b\bar{c} + ab\bar{c}$.

As the example illustrates, if a function is decomposable, then there are generally many possible choices for the decomposition functions and it seems wise to use these degrees of freedom for optimizing the costs of implementing the decomposed switching function. This problem has been recently addressed by Murgai et al. who formulated this as an input encoding problem related to the state assignment problem or the input/output encoding problem for finite state machines. It is shown [MuBr94] that the encoding problem in functional decomposition can be solved using modifications of well-known algorithms for state assignment.

The classical theory of functional decomposition mainly addresses the decomposition of single switching functions. Some recent research has addressed the decomposition of multiple-output functions. For multiple-output functions it is important to decompose in such a way that common sub-functions can be shared. This problem has been addressed by Lai et al. [LaPa96] and by Scholl and Molitor [ScMo94]. They proposed the first method for strict decomposition of multiple output functions. Wurth, Eckl and Antreich [WuEc95] suggest an approach to non-strict decomposition of multiple-output functions based on an efficient implicit enumeration algorithm and binary decision diagrams.

## 5.3 Boolean and Algebraic Methods

The automation of the design process became an important industrial factor in the 1970s and 1980s. This triggered a new wave of research in multi-level optimization aimed to overcome the complexity problems of earlier approaches (like the ones based on functional decomposition) and to develop optimization techniques capable of dealing with large industrially relevant circuits. The techniques briefly described in this section were pioneered by Brayton et al. [Bray87] and became widely accepted in both academia and industry. We will refer to these methods as *Boolean / algebraic methods*. For a more detailed description of these approaches the reader may refer to [Bray87].

### 5.3.1 Division

There are many possible views on the problem of multi-level logic minimization. As described in Section 5.2, the approaches of functional decomposition view it as a problem of decomposing a given switching function into a network of interconnected sub-functions where the individual sub-functions depend on separate sets of variables. The Boolean/algebraic techniques view the problem as a problem of performing "good" Boolean or algebraic manipulations on the representation of the given switching functions. *Division* can be considered a central issue for these manipulations.

Boolean algebra does not possess inverse operations for "multiplication" and "addition" and, hence, in the strict mathematical sense, a unique division or subtraction operation does not exist. However, in the context of optimizing switching functions, the notion of "division" is still very common and simply refers to the process of decomposing a function into a representation of the form $y = f \cdot q + r$. In the sequel, the following terminology is used. A conjunction of two Boolean expressions $f \cdot q$ is called a *Boolean product*. In the special case, if the expressions for $f$ and $q$ do not have any common variable, it is called an *algebraic product*. For example, $(a + b) \cdot (c + \bar{b})$ is a Boolean product, $(a + b) \cdot (c + d)$ on the other hand is an algebraic product. Based on this distinction algorithms for division can be classified into two categories:

> **Definition 5.1:** Given two switching functions $y$ and $f$, an operation, $(q, r) = \mathrm{DIV}\ (y, f)$, is called a *Boolean division* if it generates $q$ and $r$ such that $y = f \cdot q + r$. If $f \cdot q$ is an algebraic product the division is called *algebraic*.

Algebraic division is a special case of Boolean division. In contrast to general Boolean division, algebraic division is unique, i.e., for any $q$, $q'$ and $r$, $r'$ such that $(q, r) = \mathrm{DIV}(y, f)$ and $(q', r') = \mathrm{DIV}(y, f)$ it is $q = q'$ and $r = r'$. Circuit transformations based on the *algebraic model*, i.e., based on algebraic factorization techniques play an important role in modern logic synthesis. Being more restricted in terms of possible circuit transformations algebraic manipulations are computationally less complex than Boolean ones while they still provide satisfactory optimization results for most practical purposes. In fact, the development of efficient techniques based on the algebraic model has been an important milestone in making multi-level logic minimization feasible for very large industrial designs.

Throughout the rest of this section, it is assumed that the switching circuit is represented as a Boolean network, defined in Section 1.3.3. This leaves several choices to represent the individual network nodes. For most standard algorithms it is assumed that the switching functions are represented as sum-of-product expressions (SOPs). Therefore, in this section, unless stated otherwise, it is implicitly assumed that the considered switching functions of the Boolean network are given as SOP-expressions.

*Algebraic Division:*

The product terms in a SOP are also referred to as *cubes* and many algorithms of two- and multi-level logic minimization are based on efficient cube manipulations such as cube intersection, sum, consensus, reduction, expansion, etc. The reader may refer to [Bray84] for a detailed description of these cube set manipulations and their application to two-level logic minimization. Since it is beyond the scope of this book to formally introduce and fully describe these operations, the following discussion is based on an intuitive understanding of these operations which the reader can adopt by visualizing cubes in a three-dimensional space. Cubes represent subspaces of an $n$-dimensional Boolean space $\{0, 1\}^n$ defined by the $n$ variables of a given SOP-expression. A product term with $n$ literals determines a single vertex in this space and a product term with $s$ literals, $s \leq n$, defines an $(n\text{-}s)$-dimensional subspace or cube in $\{0, 1\}^n$ . A cube $c_1$ is said to *cover* some other cube $c_2$ , $c_2 \subseteq c_1$, if all vertices contained in the subspace defined by $c_2$ are also contained in the subspace defined by $c_1$. For example in some SOP-expression, a cube $abc$ is covered by a cube $ab$.

**Example 5.4**: Take a function $y = abc + cf + abe + de + cg + eg$ and a divisor $f = c + e$. Both, the functions and its divisor are given as a set of cubes. As the first step of an algebraic division it is checked for each cube of the divisor, which cubes of the dividend are covered by that cube.

1) Consider cube $c$: it covers $abc$, $cf$ and $cg$. This means the vertices contained in these cubes can be covered by "expanding" these cubes in the direction of $c$. The expanded cubes are $ab$, $f$ and $g$.

2) Consider cube $e$: it covers $abe$, $de$ and $eg$. The cubes expanded in the direction of $e$ are $ab$, $d$ and $g$.

3) Consider $c + e$: The vertices contained in all cubes of the quotient can be covered by taking the expanded cubes common to 1) and 2) and intersecting them with $c + e$. The expanded cubes common to 1) and 2) are $ab$ and $g$. Therefore, we obtain $y = (c + e)\ (ab + g) + cf + de$ and can identify in this expression the quotient, $q = ab + g$, and the remainder, $r = cf + de$.

The pseudo-code description of the procedure illustrated in the above example is given in Table 5.1.

Many circuit transformations that optimize a multi-level circuit can be viewed as performing "good" divisions. Good divisions, however, do not only require an efficient division method, it is just as important to identify divisors that justify the effort to attempt a division because they promise a minimization of the circuit. The heuristics to find such divisors do not always restrict the search very tightly and, therefore, a division procedure is typically called numerous times during circuit optimization. That is why researchers have dedicated much effort to make division as efficient as possible. The routine in Table 5.1, if implemented correctly, has a computational complexity of O($m$ log $m$), where $m$ is the number of cubes in $y$ and $f$. McGeer has proposed an improved method for algebraic division [GeBr88] with linear complexity if the cubes were sorted in a certain way. Such ordering needs to be performed only once as a pre-processing phase before the actual minimization starts and therefore contributes only little to the total computation time.

```
alg_division(Y, F)
/* Y: is the set of cubes of y*/
/* F: is the set of cubes d1, d2, ...dk in f */
{
  for (i := 1 to k, i :=i + 1)
  {
      Vi := set of cubes in f which are covered by di
            and with literals in di deleted;
      if (i=1)
            Q = V1;
      else
            Q := Q ∩ Vi;
  }
  R := Y \ F×Q;
  return(Q, R);
  /* Q: is the set of cubes of q*/
  /* R: is the set of cubes of r */
}
```

**Table 5.1:** Pseudo code for algebraic division procedure

*Boolean division:*

While algebraic division is very fast and can be applied even to large designs, Boolean division is a much more complex process. On the other hand, Boolean division may provide substantially better minimization results than the algebraic division and, depending on the priorities in the design, the improved minimization results may well be worth the extra effort.

> **Example 5.5:** Take $y = ac + bc + a\bar{b}$. An algebraic division using divisor $a$ results in $y = a(c + \bar{b}) + bc$. Using divisor $c$ yields $y = (a + b)c + a\bar{b}$. However, a smaller representation of this function can be obtained by a Boolean factorization. Suppose, by some good heuristic, $f = a + b$ is suggested as a divisor. Algebraic division using the algorithm of Table 5.1 generates $q = 0$ and $r = ac + bc + a\bar{b}$ and hence the representation of the function remains unchanged.

Boolean division, however, yields $q = c + \overline{b}$ and $r = 0$, i.e., we obtain $y = (a + b)(c + \overline{b})$. This is one literal less than the result found by the algebraic division.

Remember that Boolean division is not unique. For a given function and divisor different Boolean division techniques may generate different quotients and remainders and it is crucial to select a "good" solution from the set of formally correct solutions. In other words, Boolean division itself boils down to an optimization problem and many different approaches can be taken. In the following, we describe one specific Boolean division technique [Bray87]. The idea in this technique is to introduce the divisor as an additional symbolic variable into the representation of $y$ and to optimize $y$ using additional don't-care conditions.

**Example 5.5 (contd.):** Reconsider function $y = ac + bc + a\overline{b}$ . A Boolean division is to be performed for divisor $f = a + b$. This can be accomplished by formally introducing a new variable $x$ into the representation of $y$ which represents the divisor function $f$. This is illustrated in Figure 5.5 by means of the Karnaugh map of $y$.

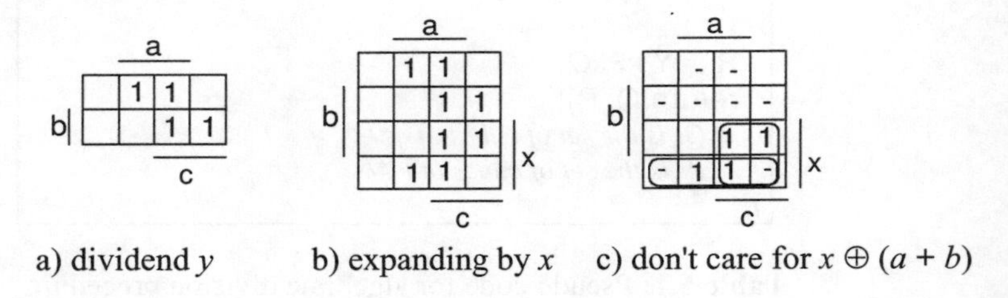


a) dividend $y$    b) expanding by $x$    c) don't care for $x \oplus (a + b)$

**Figure 5.5:** Illustration of Boolean division by Karnaugh map

Figure 5.5a shows the Karnaugh map for the dividend $y$. The division by divisor $f = a + b$ is performed as follows: first, a variable $x$ representing the divisor is formally introduced by expanding the representation of $y$. This yields the Karnaugh map of Figure 5.5b. Note that this Karnaugh map is symmetric with respect to its horizontal axis. If we tried to obtain an optimum SOP-expression from the map in Figure 5.5b, we would recognize this symmetry and derive an expression that does not depend on $x$. The trick to force $x$ into the expression is to exploit the fact that $x$ represents the divisor function and is not an independent variable. In this example, it is $x = f = a + b$. Hence, $x$ must assume the same values as $a + b$. Note that this yields don't care conditions for $y$, $DC = (a + b) \oplus x$. Entering these don't cares into the Karnaugh map of Figure 5.5b results in the map of Figure 5.5c. Extracting a minimum SOP-expression from the Karnaugh map as indicated in Figure 5.5c yields $y = xc + \overline{b}x$. In factored form we obtain $y = x(c + \overline{b}) = (a + b)(c + \overline{b})$.

The steps of a Boolean division can generally be described as follows:

a) Extend the dividend function $y$ by a variable $x$ representing the divisor $f$.
b) Form the don't care set DC = $x \oplus f = \bar{x}f + \bar{f}x$.
c) Minimize $y$ using this don't care set.

In practice, there are many possible ways to realize a Boolean division following this scheme. In MIS [Bray87], Boolean division is performed based on the two-level minimizer, ESPRESSO [Bray84]. If $y$ is represented in factored form, its representation is first changed into a SOP-form. Also the divisor must be given in SOP-form. ESPRESSO is then applied to the cube list of $y$. The cube list of $y$ is extended by a variable $x$ in a similar way as shown above for Karnaugh maps. The don't cares, DC = $x \oplus f$, are also passed to ESPRESSO and used during the two-level minimization process. The result of the minimization, if desired, is then re-factored into a factored form.

### 5.3.2 Kernel Extraction

A major task in multi-level logic minimization is to identify logic that can be shared between different output cones of the switching circuit. In the Boolean / algebraic view this involves identifying sub-expressions common to two or more expressions. While the previous sections discussed different mechanisms to perform divisions, this section addresses the problem of identifying appropriate divisors so that common logic can be shared. The common sub-expressions to be shared are termed *kernels*.

All division operations related to kernel extraction are *algebraic*. In the following the operation "/" refers to identifying the algebraic quotient of two expressions. We denote $q = y / f$ if $y = fq + r$ and $fq$ is an algebraic product.

**Definition 5.2:** A Boolean expression $y$ is called *cube-free*, if there exists no expression $q$ and no cube $c$ such that $y = cq$.

In other words, an expression is called cube-free if no cube exists that divides this expression without a remainder. Only expressions with more than one cube can be cube-free.

**Definition 5.3:** The set of *kernels* for an expression $y$ is given by $K(y) = \{y / c \mid c$ is a cube and $y / c$ is cube-free$\}$. The cube $c$ belonging to a kernel $k \in K(y)$ is called *co-kernel* of $k$.

Given an expression $y$, kernels can be computed in a straightforward way by applying the above definition. Brayton and McMullen have proposed a procedure [BrMu82] which first makes the given expression cube-free by finding the largest cube that can be factored out. The resulting cube-free factor is then used as the dividend for subsequent divisions. Proceeding in a lexicographic order, each literal of $y$ is selected and algebraic divisions are performed using these literals as divisors. The resulting quotients $q$ are checked to determine whether or not they are cube-free. If they are cube-free then they are kernels, otherwise they are made cube-free by factoring out the largest cube. This procedure can be repeated until all kernels have been generated. Note that we only have to divide by literals that occur in at least two cubes. Otherwise the resulting quotient is never cube-free and cannot yield any new kernel.

**Example 5.6:** Take function $y = acd + a\bar{b}c + b\bar{c}d + ab\bar{d}$. Dividing by the literals in $y$ yields:

$$
\begin{array}{ll}
y & \text{(cube-free)} \\
y / a := y_a = cd + \bar{b}c + b\bar{d} & \text{(cube-free)} \\
\qquad y_a / c := y_{ac} = d + \bar{b} & \text{(cube-free)} \\
y / b := y_b = \bar{c}d + a\bar{d} & \text{(cube-free)} \\
y / c := y_c = ad + a\bar{b} & \text{(not cube-free)} \\
\qquad y_c / a := y_{ca} = d + \bar{b} = y_{ac} & \\
y / d := y_d = ac + b\bar{c} & \text{(cube-free)}
\end{array}
$$

Collecting all cube-free expressions yields $K = \{y, y_a, y_b, y_d, y_{ac}\} = \{acd + a\bar{b}c + b\bar{c}d + ab\bar{d}, cd + \bar{b}c + b\bar{d}, \bar{c}d + a\bar{d}, ac + b\bar{c}, d + \bar{b}\}$. Note that $y_{ca}$ does not have to be included since $y_{ac} = y_{ca}$. To make $y_c$ cube-free the cube $a$ has been factored out to obtain $y_{ca}$. Since $a$ has already been factored out earlier to form $y_a$, $y_{ca}$ does not yield any new kernel. The reader may realize that this must always be true. In general, it can be observed [BrMu82] that if a non-cube-free expression is made cube-free by factoring out some cube, and this cube contains a literal that has already been selected for division, then the resulting expression can only contain kernels that have already been generated.

The procedure to compute kernels is given in Table 5.2. In some applications it is computationally too expensive to examine all kernels. When only subsets of all kernels are considered it is common to group the kernels according to *levels*. A kernel is of level zero if it contains no other kernel. It is of level one if it contains at least one kernel of level zero.

```
kernels(y, j)
/* y is the given expression */
/* initially j = 1 */
/* v1, v2, ...vn are the literals of y */
{
    K := ∅;
    for (i = j to n, i := i + 1)
    {
        if (literal vi is contained in at least two cubes of y)
        {
            Q := alg_division(Y, vi)    /* Y is the cube set of y */
            k := SOP-expression corresponding to Q;
        }
        if (k is cube-free)
            K := K ∪ {k};
        K := K ∪ kernels(k, i+1);
    }
    return (K);
}
```

**Table 5.2:** Pseudo-code for procedure *kernels()*

**Definition 5.4:** The subset $K^n(y)$ of kernels of a function $y$ is

$$K^n(y) := \begin{cases} \{ k \in K(y) \mid K(k) = \{k\} \} & \text{for } n = 0 \\ \{ k \in K(y) \mid \forall\, k' \in K(k), k \neq k' : k' \in K^{n-1}(y) \} & \text{otherwise.} \end{cases}$$

A kernel $k$ is of level $n$ if $k \in K^n(y)$ and $k \notin K^{n-1}(y)$.

It can be noted that the sets of lower level kernels are always subsets of the sets of higher level kernels, that is $K^0(y) \subset K^1(y) \subset \ldots \subset K(y)$.

**Example 5.6 (contd.):** Applying Definition 5.4 we obtain:

$K^2(y) = \{ acd + a\bar{b}c + b\bar{c}d + ab\bar{d}, cd + \bar{b}c + b\bar{d}, \bar{c}d + a\bar{d}, ac + b\bar{c}, d + \bar{b}\}$
$\quad = K(y)$
$K^1(y) = \{ cd + \bar{b}c + b\bar{d}, \bar{c}d + a\bar{d}, ac + b\bar{c}, d + \bar{b}\}$
$K^0(y) = \{\bar{c}d + a\bar{d}, ac + b\bar{c}, d + \bar{b}\}$

Hence, $\bar{c}d + a\bar{d}$, $ac + b\bar{c}$ and $d + \bar{b}$ are kernels of level zero, $cd + \bar{b}c + b\bar{d}$ is a kernel of level one and $y = acd + a\bar{b}c + b\bar{c}d + ab\bar{d}$ is the only kernel of level two.

Often it is desirable to only consider kernels of level zero. Therefore, modifications to the algorithm in Table 5.2 have been suggested [Bray87] that allow to quickly extract one or all kernels of level zero. More recently, a fast and powerful kernel extraction method to identify single- and double cube divisors has been proposed by Rajski and Vasudevamurthy [RaCo90].

### 5.3.3 Optimization Strategies

Previous sections have described methods to manipulate Boolean expressions by division and an algorithm to extract certain sub-expressions called kernels. It remains to be shown how these techniques can be combined in a general synthesis framework for use in specific minimization strategies. The described minimization strategies correspond to commands in a public domain synthesis tool SIS (contains MIS). A typical optimization run invokes a series of operations in the Boolean network. In SIS user-defined *scripts* specify what operations shall be performed and in what order. This section describes in a simplified manner the important operations of *factoring*, *extraction*, *resubstitution* and *simplification*.

*1) Factoring*

Factoring is the process of transforming a SOP-expression into a factored form with more than two levels. For example, the SOP-expression $y = ace + bce + de + f$ can be expressed in a factored form as $y = ((a + b)c + d)e + f$ to save three literals. Factoring is used in multi-level synthesis to simplify the representation of a node in the Boolean network. An expression is factored by performing a sequence of divisions by well-selected divisors. In most applications the divisors are selected from the set of kernels and the division is algebraic.

The procedure of factoring is given in Table 5.3. Note that *factor()* is only used to simplify the representation of a node in the Boolean network. No nodes are deleted and no new nodes are introduced. Factoring of the nodes in a network typically has to be conducted many times during the optimization procedure because only a factored representation of the nodes gives a realistic estimate on the actual area requirements of the circuit to be synthesized.

```
factor(y)
{
    k := selected kernel generated by kernel(y, 1);
    if ( k = ∅)
        return (y);
    (q, r) := division(y, k);  /* this can be algebraic or Boolean division */
    return (factor(k) · factor(q) + factor(r);
}
```

**Table 5.3:** Pseudo-code for procedure *factor()*

*2) Extraction*

Extraction is the process to identify kernels or cubes that are common to two or more expressions and can serve as common divisors for several expressions. Extraction is therefore an important instrument to introduce sharing of logic between different circuit cones, thus minimizing the circuit. Given the expressions of several nodes and a common divisor, the operation of extraction introduces a new node for the divisor which fans out to the quotients resulting from the given expressions by division. An important issue is therefore to identify particularly good common divisors yielding maximum simplification after division for a maximum number of expressions. Efficient solutions to this problem are supported by formulating it as a rectangle covering problem as proposed by Rudell [Rude89].

**Example 5.7** [DeMi94]: Extraction is demonstrated by the example of Figure 5.6.

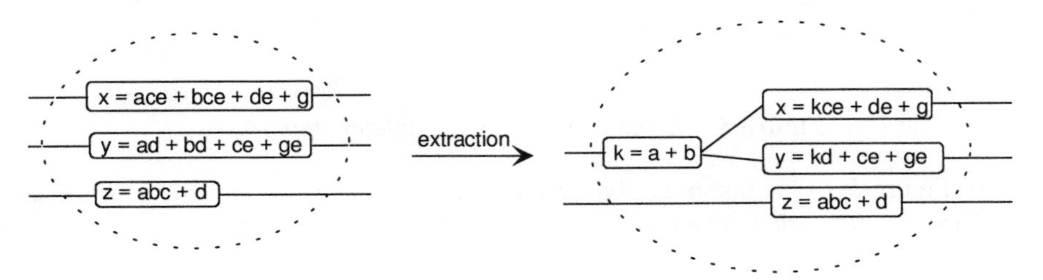


**Figure 5.6** Extraction in a Boolean network

The following procedure for kernel extraction lacks sophistication but illustrates the basic steps performed.

a) compute all kernels:
   $K(x) = \{a + b, ac + bc + d, ace + bce + de + g\}$
   $K(y) = \{a + b, c + g, ad + bd + ce + ge\}$
   $K(z) = \{abc + d\}$
b) determine kernel intersections: $a + b$ is a common kernel

c) create new network node $k$ with $k = a + b$
d) replace all expressions $v$ containing kernel $k$ by $k \cdot v / k + r$, where $r$ is the remainder of the division: $x = kce + de + g, y = kd + ce + ge$.

It is generally too complex to determine kernel intersections in a brute force manner like in the above example. More advanced techniques to examine kernel intersections for extraction can be found in [Bray87].

*3) Resubstitution*

Another important network manipulation is *substitution* or *resubstitution*. Resubstitution does not add or remove any nodes in the Boolean network, however it introduces new branches. Resubstitution is the process of dividing the function of a node in the network by the function of some other node in that network. Literals can then often be saved if the dividend is replaced by a smaller quotient in conjunction with the divisor variable. This is illustrated in the following example.

**Example 5.8:** Resubstitution is illustrated in Figure 5.7. The network is simplified by taking the function of some node in the network as divisor for some other node.

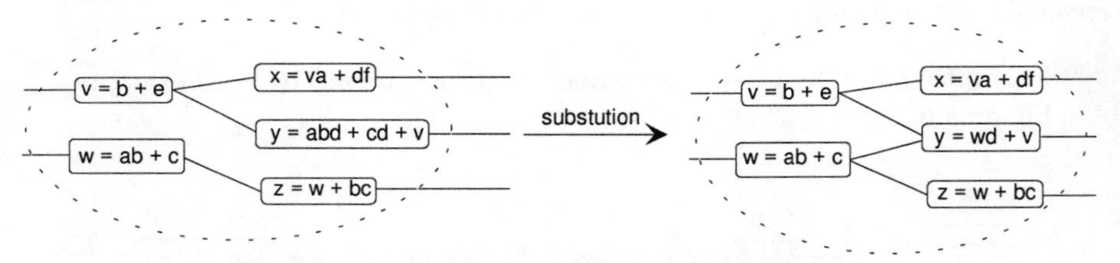


**Figure 5.7:** Resubstitution in a Boolean network

In Figure 5.7 $w$ is taken as divisor for $y$. The division $y/w$ yields $y = wd + u$. Then the network can be re-structured as shown.

Depending on the type of division performed the resubstitution is called algebraic or Boolean. In practice, both types are employed. Algebraic resubstitution is discussed in [Bray87]. In BOLD [HaJa88] Hachtel et al. suggested a Boolean resubstitution technique based on ESPRESSO. Heuristics are needed to decide whether or not attempting a resubstitution at some node $a$ using some divisor node $b$ is worthwhile. For algebraic resubstitution there are some simple checks [Bray84] to reduce the number of possible divisors to be tried. This is more difficult for Boolean resubstitution. For Boolean resubstitution it is much harder to predict its promise and this is particularly delicate because of the significant computational costs associated with Boolean operations. A new, strong heuristic for selecting good divisors for substitution will be discussed in Section 5.6.2. It has been shown in [KuMe94] that certain types of logic implications can give helpful information in this context. In fact, if an *indirect* implication, defined in Section 2.3, can be identified between two nodes of the network, then it is a strong indication that a minimization can be obtained from a substitution involving these two nodes. This will be further discussed in Section 5.6.2.

*4) Simplification*

Another important operation in a Boolean network is node *simplification*. This leaves the structure of the network unchanged while locally minimizing the representation of each node in the network. Two-level minimization techniques and factoring can be used for this purpose. However, in order to obtain more degrees of freedom for node simplification it is important to calculate don't care conditions that apply at each node. Computing don't care conditions at internal nodes in the circuit is of great importance not only for simplification but for all *Boolean* techniques. Remember that Boolean division as described in the previous section is based on logic minimization, e.g., by ESPRESSO. As is well-known a stronger minimization effect can be observed if there are don't care conditions that can be passed on to the optimizer. Therefore, computing internal don't care conditions to enhance Boolean manipulations will be briefly considered in the next section.

### 5.3.4 Internal Don't Cares

Not all combinations of binary value assignments can generally occur at the input variables of an internal node in a Boolean network. Since these variables are not *independent*, some combinations of value assignments are "filtered out" by the preceding logic. Clearly, the function of the considered node only needs to be specified for those value assignments that can actually occur and consequently all other combinations of value assignments represent a *don't care set*. Such don't care conditions are also called *controllability* don't cares.

Let $s$ be an internal variable associated with a switching function $f(\boldsymbol{x})$. The variable $s$ can only assume logic values that are actually produced by function $f$, i.e., any combination of value assignments in the circuit for which $s \oplus f(\boldsymbol{x}) = 1$ cannot occur and hence represents a controllability don't care set. The set CDC of all controllability don't cares in the circuit is given by

$$\text{CDC} = \sum_i \; (s_i \oplus f_i(\mathbf{x})), \qquad \text{(Eq. 5.2)}$$

where $s_i$ are the variable names associated with the functions $f_i(\mathbf{x})$ of all nodes $i$ in the network. CDC can be understood as a Boolean function over all external and internal variables of the network which evaluates to one if the corresponding combination of value assignments in the network cannot occur.

Controllability don't cares are not the only degrees of freedom that can be taken into account. Often, the value of an internal logic function cannot be *observed* at any of the circuit outputs and therefore, in such situations the value of the function need not be specified. This leads to the so called *observability don't cares*. For a given situation of value assignments in the Boolean network an internal node $x$ is *observable* at an output $y$ of the network iff any change in the value of $x$ changes the value at $y$. This can be described by the notion of *Boolean difference*: $\partial y/\partial x = y|_{x=1} \oplus y|_{x=0}$ evaluates to one for all situations of value assignments in which $x$ is observable at $y$. The set ODC($x$) of all observability don't cares for an internal variable $x$ is given by

$$\text{ODC}(x) = \prod_i \; y_i|_{x=1} \overline{\oplus} \; y_i|_{x=0} \qquad \text{(Eq. 5.3)}$$

where $y_i$ are the primary outputs of the Boolean network.

Computing these don't care functions is generally not an easy task. The number of don't care conditions is often huge and may require excessive amounts of memory for its representation. Therefore, only partial don't care information can sometimes be computed. Significant progress has been made towards solving this problem by applying techniques originally developed for image computation in finite state machines [SaBr91] and by representing the don't care functions using binary decision diagrams. For a more detailed description of these methods the reader may refer to [DeMi94], [HaSo96].

## 5.4 Global Flow

The *global flow* approch for circuit minimization adopts techniques from compiler optimization. It is a fast optimization procedure employed in IBM's logic synthesis system LSS and can be applied to very large designs. It was originally proposed by Trevillyan et al. [TrJo86], [BeTr91] and then extended or modified [BrSe88], [YuNa95]. In this section we mostly follow the notation and description of the global flow procedure as in [BrSe88].

Global flow optimization is performed by repeatedly performing the following two steps. The first step calculates information about the circuit in terms of *implications*. In the second step this information is used to transform and optimize the circuit. This methodology is related to the approach to be presented in Section 5.6.

In global flow, information about the circuit is collected by calculating approximations to so called *forcing sets* defined as follows:

**Definition 5.5:** For each signal $x$ in the circuit four sets of signals $F_{VW}(x)$ with $V, W \in \{0, 1\}$ called *forcing sets* are determined such that

$$F_{VW}(x) \equiv \{ s: x = V \Rightarrow s = W\}, \qquad V, W \in \{0, 1\}$$

For example, if $y \in F_{10}(x)$ then the implication $x = 1 \Rightarrow y = 0$ is true. Calculating the complete forcing sets is a complex problem. In principle, if given enough time, they could be calculated using the recursive learning procedure of Section 3.2.1. In practice only approximations for the forcing sets can be determined. One possibility to obtain good approximations would be to restrict the recursion depth for recursive learning. A different approach is taken in [BeTr91]. There, subsets of the forcing sets called *controlling sets* are calculated simultaneously for all nodes in the network by recurrence relations. In the following, we restrict considerations to the controlling sets $C_{10}(x)$ and $C_{11}(x)$ as approximations for $F_{10}(x)$ and $F_{11}(x)$, respectively. Further, it is assumed that the circuit only consists of NOR gates. All other cases follow analogously.

$$\begin{aligned} C_{10}(x) = C_{10}(x) \cup & \\ & \{s: \exists(y, s)[\, y \in C_{11}(x)]\} \cup \\ & \{s: \exists(s, y)[\, y \in C_{11}(x)]\} \cup \\ & \{s: \text{x} \in C_{10}(s)\} \end{aligned} \qquad \text{(Eq. 5.4 )}$$

$$\begin{aligned} C_{11}(x) = C_{11}(x) \cup & \\ & \{s: \forall(y, s)[y \in C_{10}(x)]\} \cup \\ & \{s: \exists(s, y), y \in C_{10}(x), \forall(t, y)[t \neq s \Rightarrow t \in C_{10}(x)]\} \cup \\ & \{x\} \end{aligned} \qquad \text{(Eq. 5.5)}$$

In these recurrence equations $(y, s)$ is a pair of signals such that $y$ is the input signal of a gate with output $s$. The sets $C_{10}(x)$ and $C_{11}(x)$ are determined by iterating through the above equations until no more changes occur in any of the sets. Consider the second clause in Eq. 5.4. It expresses that $C_{10}$ must contain all signals $s$ that are outputs of a NOR gate with at least one input assuming the logic value 1. This is because such signals always assume the value 0. The third clause expresses that all inputs $s$ of a gate must be contained in $C_{10}$ if the output $y$ of the gate has the value 1. The forth clause takes into account contraposition. If $s = 1 \Rightarrow x = 0$ then $x = 1 \Rightarrow s = 0$ is also true.

We note that the controlling sets contain all *direct* implications as well as those *indirect* implications that can be obtained from direct implications by contraposition. Thus, solving the above recurrence relations leads to identifying the same set of implications as can be calculated by *static learning*, described in Section 2.7.

Similarly, like in the approach to be presented in Section 5.6 the collected implications are used to transform the circuit according to the following theorem.

**Theorem 5.3:** Let $f$ and $y$ be nodes of a combinational network $N$. Then, the following transformations on the combinational network are valid:

1) If $y \in F_{11}(x)$, then replace $y$ by $x + y$.
2) If $y \in F_{10}(x)$, then replace $y$ by $\overline{x} \cdot y$.

The transformations in Theorem 5.3 add extra logic to the circuit. Area savings can be achieved by subsequently determining other logic that can be removed.

**Definition 5.6:** In a combinational network $N$ the *1-frontier* of a signal $x$ is defined as the set of signals $s$ such that

1) $s \in C_{1V}(x)$, $V \in \{0, 1\}$,
2) there is a path $(s, j_1, j_2, ....po)$, where $po$ is a primary output of $N$, such that no $j_k$ is in $C_{1V}$,
3) $s$ is in the transitive fanout of $x$.

**Definition 5.7:** $G_1(x)$ is a graph consisting of nodes $j \in C_{1V}(x)$, $V \in \{0, 1\}$, $j$ in the transitive fanout of $x$, and edges $(j, k)$ where $k \in C_{1V}(x)$ and $j$ is an immediate predecessor of $k$. In addition, add edges $(j, k)$ if there is a path from $j$ to $k$, where the only nodes along the path in $C_{1V}$ are $j$ and $k$.

Definition 5.7 defines an implication graph that describes the "logical flow" of the value assignment 1 at signal $x$ in forward direction. For $x = 0$ the graph is analogous. The additional edges in Definition 5.7 are needed to cover the cases where forward implications exist at a node but do not result from the values at the immediate predecessors. This can happen for indirect forward implications like in Figure 3.2.

**Definition 5.8:** A *1-cut-set* of a signal $x$ is a set of signals in any of the $C_{1V}(x)$, $V \in \{0, 1\}$ which separates $x$ from its 1-frontier in the implication graph $G_1(x)$.

As a special case the elements of the 1-frontier represent a 1-cut-set.

**Theorem 5.4:** If all gates of a chosen 1-cut-set of $x$ are transformed according to Theorem 5.3, then any connection of $x$ to a gate $k \in G_1$, k $\notin$ 1-cutset, can be replaced by a constant zero without affecting the function of the circuit.

```
/*     this procedure takes as parameter a gate netlist
       description N of the circuit */

global_flow_optimize(N)
{
  /* determine implications */
  compute the sets Cvw(x) for all nodes x in N;

  /* perform circuit transformations */
  for (all nodes x in N and every value V ∈ {0, 1})
  {
       using Cvw(x) find the V-frontier of x;
       find a minimum weighted V-cutset of signals
           separating x and its V-frontier;

       /* rearrange connections*/
       add connections from x  to gates in V-frontier as by Theorem 5.3;
       remove connections of x as by Theorem 5.4;

       if (area of N has improved)
       /* controlling sets must be updated if circuit is modified */
           re-compute Cvw(x) for all x in the network;
       else
           reverse circuit transformation;
  }
  return N;
}
```

**Table 5.4:** Pseudo-code description of global flow optimization

For 0-frontiers and 0-cutsets the tranformations are analogous. Global flow optimization is the process of rearranging the connections of a wire $x$ without changing its associated 1-frontier (0-frontier). The circuit is modifed by first adding connections from $x$ to selected gates according to Theorem 5.3 and then by removing other connections of $x$ according to Theorem 5.4. Not all of these transformations necessarily result in an optimization of the circuit. To obtain good optimization results it is very important to choose appropriate cut sets of $x$. For this purpose a min-cut-procedure is applied to the implication graph $G$(x) [BeTr91].

Table 5.4 summarizes the general procedure for global flow optimization. Good heuristics are needed to find appropriate cuts through the implication graph. Note that the controlling sets need to be updated each time the circuit has been modified. This problem is similar to that occurs in methods exploiting internal don't care sets and has been addressed by "invariance research" [GeBr89]. For global flow an incremental update procedure has been suggested [BrSe88].

**Example 5.9:** Figure 5.9a shows a circuit example to be optimized by the global flow approach. First the controlling sets are calculated by the recurrence equations. Assume $x = 1$ and we observe the forward implications of this assignment. The forward implication yields $y = 1$, i.e., we obtain $y \in C_{11}(x)$.

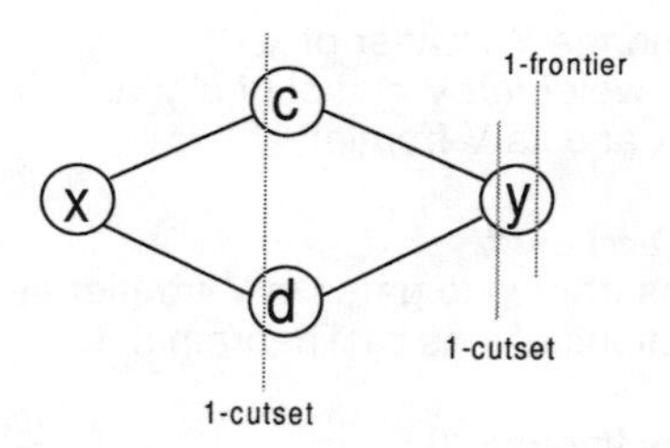


**Figure 5.8:** Graph $G_1(x)$ with 1-frontier and two possible 1-cutsets

The forward implications do not reach any further than the node $y$ since it is a primary ouput. Hence, node $y$ is element of the 1-frontier of $x$. Figure 5.8 shows the implication graph $G_1(x)$. The 1-frontier only consists of node $y$ and can also be considered as possible 1-cutset. This is indicated in Figure 5.8. Another possible cutset is also shown. Heuristically, the cutset should be small and applying the min-cut procedure of [BeTr91] to the shown implication graph returns node $y$ as 1-cutset.

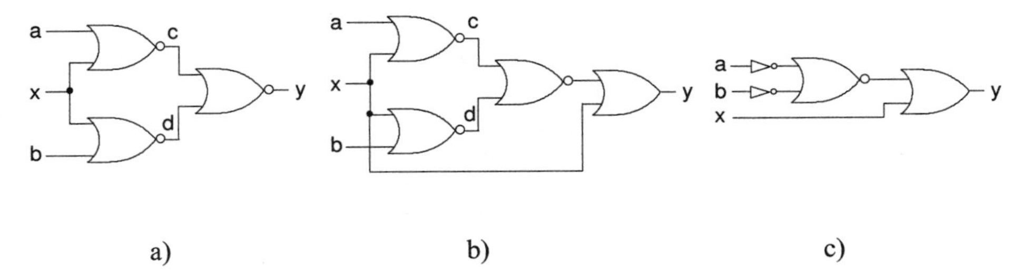


**Figure 5.9:** Circuit optimization by global flow

As described in Table 5.4 the transformation of Theorem 5.3 is performed to obtain the circuit of Figure 5.9b. Then, Theorem 5.4 is applied, i.e., all connections emerging from $x$ that do not feed a gate of the chosen 1-cutset are set to 0. In this example, these are the connections of $x$ to the gates with outputs $c$ and $d$. The connection from $x$ to $y$ must not be touched because $y$ belongs to the 1-cutset. The nodes with constant values can be removed so that the NOR gates with outputs $c$ and $d$ degenerate to inverters and the circuit of Figure 5.9c is obtained.

## 5.5 Redundancy Elimination

Redundancy elimination is a process based on extensions to any conventional test generation procedure for single stuck-at faults. By eliminating circuit elements related to untestable faults redundancy elimination makes a circuit 100% testable with respect to single stuck-at faults.

The redundancy elimination process is easily understood by observing the following: suppose a single stuck-at-$V$ fault, $V \in \{0, 1\}$, cannot be tested. This means that no set of value assignments at the inputs of the circuit exists for which this fault causes a wrong response at any of the outputs. Consequently, if the corresponding signal is set to the constant value $V$ that does not affect the function of the circuit. Hence, the untestable fault can be removed from the combinational network by setting this signal to $V$ and simplifying the circuit accordingly. Figure 5.10 shows a flowchart that describes more specifically the process of removing a given untestable fault.

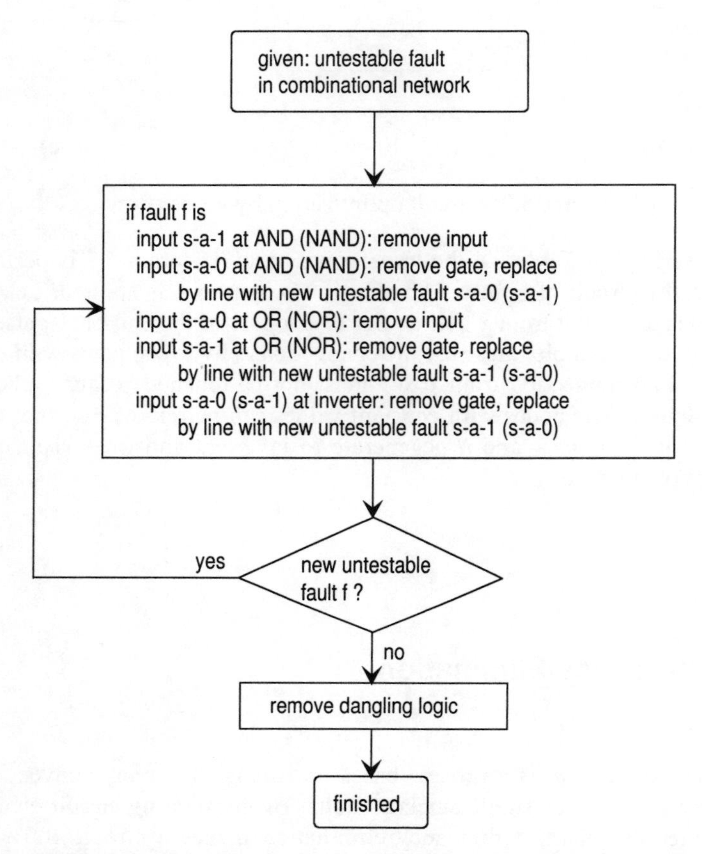


**Figure 5.10:** Procedure to remove logic related to a given untestable fault

**Example 5.10:** Figure 5.11a shows an example of a combinational circuit containing an untestable fault. This fault can be removed by the procedure of Figure 5.10. Figure 5.11b shows an intermediate step that occurs after the procedure has passed only once through the loop shown in Figure 5.10. The final circuit is shown in Figure 5.11c.

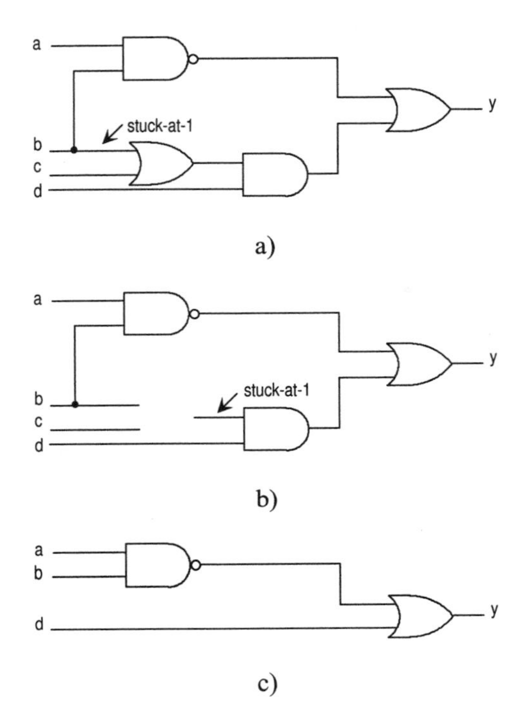


**Figure 5.11:** Removing untestable fault in combinational circuitry

The general procedure of removing all untestable faults is analogous to test generation and is shown in Figure 5.12. Just like in test generation, fault simulation techniques can be used to speed up the process. The only difference to a standard test generation technique is that if an untestable fault is encountered an additional procedure is called to remove the related logic from the network. Importantly, only one redundant fault can be removed at a time. Removing an untestable fault can create new untestable faults so that the whole procedure has to iterate until no new redundancies can be found.

Figure 5.12 shows a flowchart of the standard approach to redundancy removal based on ATPG. This procedure represents an important sub-routine for the logic optimization technique to be described in Section 5.6. A more sophisticated approach for ATPG-based redundancy removal that requires only one pass through the circuit has been suggested [AbIy92].

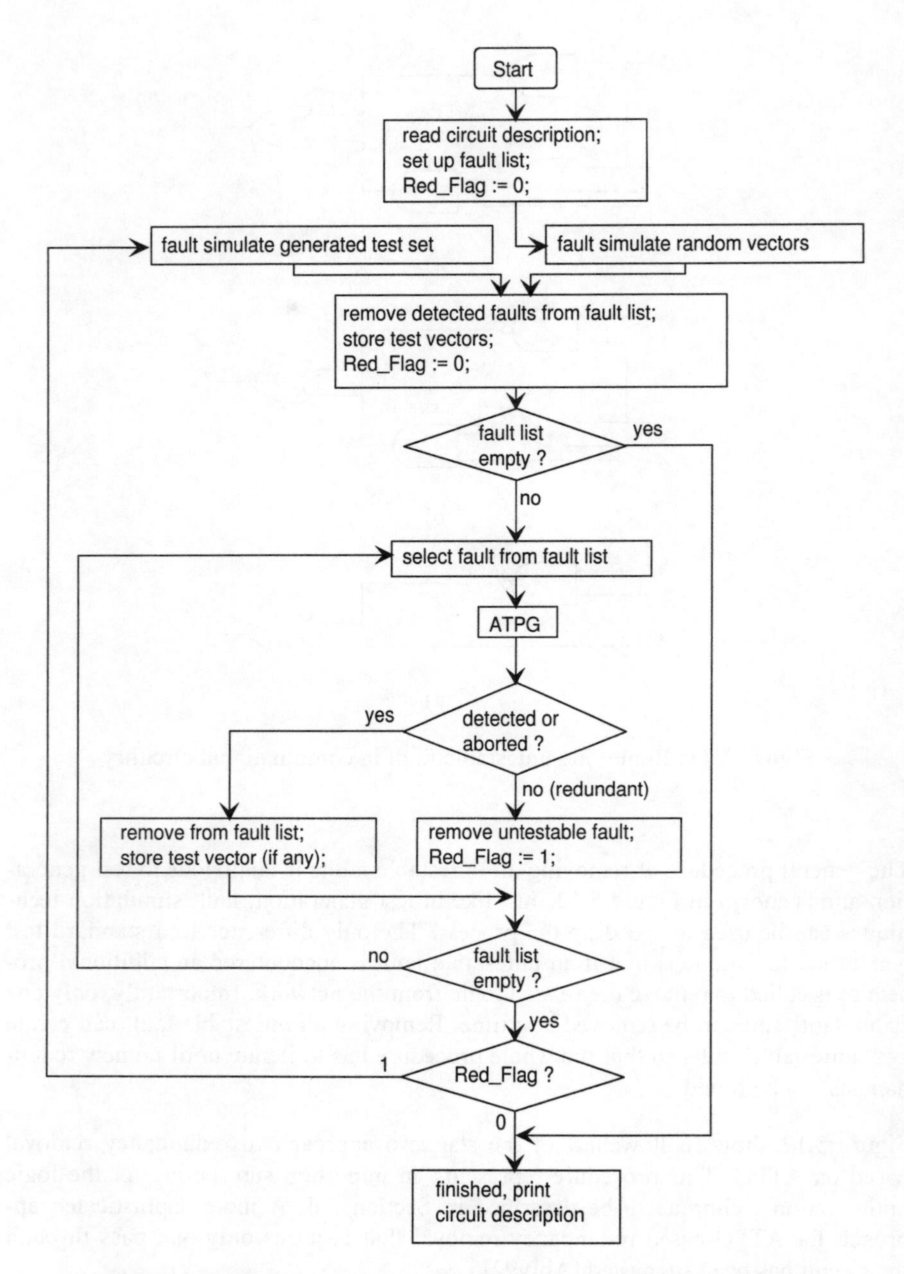


**Figure 5.12:** Redundancy elimination in combinational network

If a circuit is 100% testable for single stuck-at faults it is impossible to remove any single gate or input of a gate. This is related to the notions of *prime* and *irredundant* used in two-level circuit theory (see Section 1.3.2). Obviously, a circuit in SOP-form is prime and irredundant if and only if it is 100% testable for single stuck-at faults. The notions of prime and irredundant can also be applied to multi-level combinational networks [BaBr88]:

> **Definition 5.9:** A combinational network is called *irredundant* if no single gate can be removed, it is called *prime* if no single input of a gate can be removed without changing the function of the circuit.

In these terms, redundancy elimination represents an ATPG-based procedure to make a combinational network prime and irredundant.

## 5.6 Implication-Based Methods

Most common synthesis procedures divide the synthesis process into a technology-independent minimization phase and a *cell-binding* procedure which maps the design to a specific target technology (*technology mapping*). However, the strict separation of logic minimization from the specific technology information can sometimes be of disadvantage since the powerful concepts for deriving circuit transformations cannot be properly oriented. An important practical motivation of the approaches described in this section is to work towards general logic minimization techniques which operate directly on the structural gate netlist description of the circuit so that the specific technological information of the given gate library can guide the optimization process. This is one of the reasons why the techniques to be described are based on ATPG and on the AND/OR reasoning methods of Chapters 3 and 4 that operate directly on a gate-level circuit description.

This section describes a new paradigm to manipulate and optimize multi-level combinational networks. It is based on exploiting implications and implicants determined by the methods of Chapters 3 and 4. Implicants have always played an important role in two-level minimization. For two-level circuits an exact minimization algorithm has already been proposed long ago and is known as the Quine-McCluskey method [Quin52], [McCl56]. To minimize a Boolean function, the Quine-McCluskey method first generates all prime implicants and then identifies a subset of them that covers the function at a minimal cost. Exact minimization may require too much computation for large designs and therefore heuristic methods have been developed as in ESPRESSO [Bray84]. In simplistic terms, a heuristic minimization procedure for two-level circuits can be described as follows:

> For a given cover (see Section 1.3.2) of the function to be minimized:
>
> Step 1: add circuitry that does not change the function of the circuit, e.g. make prime implicants non-prime or add new implicants making the cover redundant
>
> Step 2: make the cover prime and irredundant (guided by appropriate heuristics), if minimization result is not satisfactory go to Step 1, otherwise output current cover as minimized cover.

The idea is to "shake up" the circuit by adding redundant logic which causes also other parts of the circuit to become redundant. Removing the redundant logic in appropriate order in Step 2 can lead to a different prime and irredundant implementation representing a cheaper cover of the function.

In this section we describe a heuristic procedure for *multi-level* minimization relying on a similar heuristic approach. For Step 1, we use the methods of Chapters 3 and 4. For a given node in the combinational network implicants are determined and are added to the cover for that node, hence making the circuit redundant. There is usually a huge choice of implicants and it is very important to select the right ones in order to obtain good optimization results. We will show in Section 5.6.2 that topological properties of AND/OR reasoning trees can provide good guidance. For Step 2 in the above procedure, we use ATPG-based redundancy elimination as described in Section 5.5.

How can this general procedure be related to the Boolean/algebraic techniques [Bray87] that have become widely accepted? As summarized in Section 5.3, the Boolean/algebraic techniques describe circuit transformations as divisions exploiting internal don't cares. In order to better understand the relationship between the optimization procedure based on implicants and ATPG (to be described in the following sections) and the conventional concepts as in Section 5.3, the next section mathematically describes circuit transformations by *orthonormal expansions*. This description shows the internal don't cares that are used in each transformation and allows us to easily relate our approach to both, the general two-step minimization procedure given above and the Boolean division techniques of Section 5.3.

We are not able to provide an exact multi-level logic minimization method that is tractable for circuits of large size. However, in the next section we will theoretically prove that circuit transformations derived by implicants generalized for multi-level networks as introduced in Section 4.3, in combination with ATPG-based redundancy elimination can in fact perform arbitrary manipulations in a combinational network. This means that our extended notion of implicants is sufficient to describe the basic "logic pieces" for manipulating not only two-level circuits but also multi-level circuits: if the implicant-based manipulations are guided appropriately, then, in principle, iterating through the above two-step procedure does not only lead to an exact minimum solution for two-level circuits but also for multi-level circuits.

### 5.6.1 Transforming Combinational Circuits by ATPG and Implications

The first important contribution using an ATPG approach to multi-level optimization has been presented by Entrena and Cheng [EnCh93]. They introduce an extension to redundancy removal which is based on adding and removing connections in the circuit. Further improvements of this approach have been reported in [ChMa94]. The approach to be described here is based on [KuMe94] and uses related mechanisms. However, there are some important differences between the approaches of [EnCh94], [ChMa94] and the method developed here: the technique proposed here can perform more general transformations in the circuit. Circuit transformations are identified by implications and the optimization process is guided by topological properties of AND/OR reasoning graphs.

Other related work includes the methods of Rohfleisch and Brglez [RoBr94] whose method is based on the introduction of *permissible bridges* and it is shown that such approach is also applicable after technology mapping. Applications to timing optimization have been examined in [RoWu95].

Let the circuit be represented by a combinational network $C$ as defined in Section 1.3.3 with $n$ primary inputs and $m$ primary outputs. Furthermore, assume that there are no *external* don't cares, the function of the combinational network $C(\boldsymbol{x})$: $B^n_2 \to B^m_2$ with $B_2 = \{0, 1\}$ is completely specified. An extension of the methods using external don't cares is possible, but will not be considered here.

Two combinational networks C and C' are called *equivalent,* denoted C = C', if they implement the same function $C(\boldsymbol{x})$: $B^n_2 \to B^m_2$ with $B_2 = \{0, 1\}$. They are called *structurally identical* or simply *identical* if there exists a one-to-one mapping between the nodes of C and C', such that for every node $y_i \in$ C there is a $y_i' \in$ C' and vice versa, where $y_i$ and $y_i'$ implement the same function. We denote identical combinational networks by $C \equiv C'$.

A basic technique to describe manipulations of switching functions is the well-known Shannon expansion, already introduced in Section 1.3.2. Let $y$ be a Boolean function of $n$ variables $x_1...x_n$. The Shannon expansion for $y$ with respect to $x_i$ is given by:

$$y(x) = y(x_1,... x_n) = x_i \cdot y(x_1,.. x_i = 1,.., x_n) + \overline{x_i} \cdot y(x_1,.. x_i = 0,.. x_n) \quad \text{(Eq. 5.6)}$$

Shannon's expansion can be understood as a special case of an orthonormal expansion [Brow90] given by

$$y(x) = \sum_{i=1}^{k} f_i(x) \cdot y_i(x)|_{fi}$$

where the functions $f_i(x)$, $i = 1, 2,..., k$ represent an orthonormal basis, i.e.,

i) $f_i(x) \cdot f_j(x) = 0, \forall i \neq j \in \{1, 2, ...k\}$ and

ii) $\sum_{i=1}^{k} f_i(x) = 1.$

The functions $y_i(x)|_{fi} : B^n{}_2 \rightarrow B_2$, $B_2 = \{0, 1\}$ are called the *(generalized) cofactors* with respect to the functions $f_i(x)$. Shannon's expansion is the special case of the above expansion where

$k = 2$,
$f_1(x) = x$ and
$f_2(x) = \overline{x}$,

and $x$ is some variable of $y$. Next we consider another special case of the above orthonormal expansion where, generalizing Shannon's expansion, we choose

$k = 2$,
$f_1(x) = f(x)$ and
$f_2(x) = \overline{f}(x)$,

where $f(x)$ is some arbitrary Boolean function $f(x)$: $B^n{}_2 \rightarrow B_2$, $B_2 = \{0, 1\}$. This means we obtain an expansion given by the following equation:

$$y(x) = f(x) \cdot y(x)|_{f(x)=1} + \overline{f}(x) \cdot y(x)|_{f(x)=0} \quad \text{(Eq. 5.7)}$$

(short notation: $y = f \cdot y|_1 + \overline{f} \cdot y|_0$ )

The terms $y(x)|_{f(x)=1}$ and $y(x)|_{f(x)=0}$ denote the cofactors of this expansion. In the special case of Shannon's expansion, as is well known, there is a simple rule on how to choose cofactors. As by Eq. 5.6 the cofactors are usually chosen by *restricting* the original function with respect to the corresponding variable. We obtain the cofactor for $y$ with respect to a variable $x$ by setting $x = 1$ in the expression for $y$.

Similarly, the cofactor for $\bar{x}$ results when we set $x = 0$. Note that there is no such simple rule in the more general case of Eq. 5.7.

Let the cofactors be denoted $y(\boldsymbol{x})|_{f(x)=V}$, with $V \in \{0, 1\}$. Further let $y^*(\boldsymbol{x})|_{f(x)=V}$ denote an incompletely specified function $B^n{}_2 \rightarrow B_3$, $B_3 = \{0, 1, X\}$. The cofactors in Eq. 5.7 must be chosen such that the following equation holds:

$$y(\boldsymbol{x})|_{f(x)=V} \supseteq y^*(\boldsymbol{x})|_{f(x)=V} := \begin{cases} y(x), \text{ if } f(x) = V \\ \text{X (don't care) otherwise} \end{cases} \qquad \text{(Eq. 5.8)}$$

It is easy to verify the validity of Eq. 5.8. Assume the truth table of $y$ is divided into two parts such that $f$ is false for all rows in the first part and true for all rows in the second part. If we first consider the part of the truth table of $y$ for which $f$ is true, we can set $y$ to don't care value for all rows in which $f$ is false. This means that the cofactor function must only have the same value as $y$ in those rows where $y$ is not don't care. Therefore, any valid cofactor for the expansion of Eq. 5.7 *covers* (denoted '$\supseteq$' ) the incompletely specified function $y^*(\boldsymbol{x})|_{f(x)=1}$ as given by Eq. 5.8. This first part of the function is described by the expression $f(\boldsymbol{x}) \cdot y(\boldsymbol{x})|_{f(x)=1}$. In the second part we have those rows of the truth table for which $f$ is false and we obtain $\overline{f}(\boldsymbol{x}) \cdot y(\boldsymbol{x})|_{f(x)=0}$.

Eq. 5.7 is the basis of our approach to transforming a combinational network. In order to relate our approach to the Boolean algebraic techniques described in Section 5.3 we can refer to function $f(x)$ as *divisor* of $y(x)$. Similarly, $y(\boldsymbol{x})|_{f(x)=1}$ can be referred to as *quotient* and the expression $\overline{f}(\boldsymbol{x}) \cdot y(\boldsymbol{x})|_{f(x)=0}$ represents the *remainder* of the division. It is interesting to observe that the combined don't care sets of the two cofactors in Eq. 5.8 are identical to the don't care set that is given to a minimization algorithm if Boolean division is performed as described in Section 5.3.1. Similarly, like in the formulation of [Bray87], the main issue in the approach to be described is to find appropriate (divisor) functions $f(x)$ such that the internally created don't cares as given by Eq. 5.8 provide "degrees of freedom" in the combinational network which can be exploited to minimize its area.

The result of the above expansion or Boolean division depends on how the don't cares are used in order to minimize the circuit. Remember that Boolean division is not unique. In [Bray87], the don't cares are explicitly calculated to be used in optimization by ESPRESSO. The approach to be described [KuMe94], proceeds in a different way and uses a *test generator* to determine the cofactors in the above expansion. As already observed by Brand [Bran83], circuitry tends to have an increased number of untestable single stuck-at faults if it is not properly optimized with respect to a given don't care set. This also suggests that the don't cares introduced by the sub-optimal circuitry created by the expansion of Eq. 5.7 may cause untestable stuck-at faults which can be removed by the standard procedure of *re-*

*dundancy elimination*. In fact, redundancy elimination is a simple way to minimize the circuit with respect to don't care conditions. Note that redundancy elimination does not require any *explicit* knowledge about don't care sets. Throughout this section, transformations are examined that create *internal* don't cares. However, these don't care conditions are not explicitly calculated or represented. They are only considered in our theoretical analysis to find the causes of the redundant faults.

**Example 5.11:** To illustrate how don't cares given by Eq. 5.8 lead to untestable stuck-at faults consider Shannon's expansion as an example, i.e., take the special case where the divisor $f(x)$ is some variable $x_i$. Note that the original function $y$ is a possible cover for both $y^*(\mathbf{x})|_{f(x)=1}$ and $y^*(\mathbf{x})|_{f(x)=0}$ so that we can form the equation $y = x_i \cdot y + \overline{x_i} \cdot y$. Taking the original function as (trivial) cofactor is certainly not the best possible choice when trying to save literals. This is because the don't care conditions given by Eq. 5.8 are not used in order to simplify the representation. If implemented as combinational circuit this leads to untestable stuck-at faults as shown for the example in Figure 5.13.

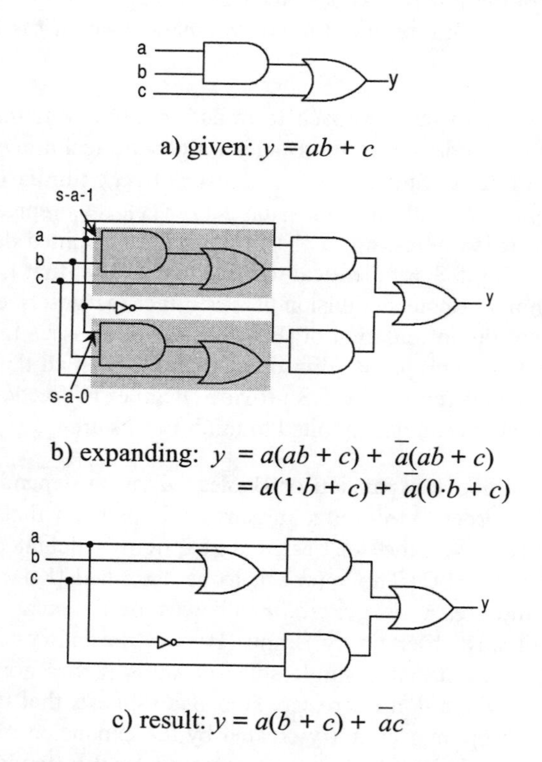


**Figure 5.13:** Shannon's expansion by ATPG

Consider Figure 5.13b. The fact that the cofactors (shaded gray) are not optimized with respect to the don't cares existing according to Eq. 5.8 explains why untestable stuck-at faults are obtained in the cofactors. By ATPG it is determined that $a$ stuck-at-1 and $a$ stuck-at-0 in respective cofactors are untestable and can be removed by setting $a$ to constant 1 and 0, respectively. Redundancy removal in this case obviously corresponds to setting $x_i$ to 1 or 0 in the respective cofactors of Eq. 5.6.

By viewing redundancy elimination as a method to set signals in cofactors to constant values, we have just described an ATPG-based method to perform a Shannon expansion. Clearly, it is not sensible to use a test generator just to prove that $x_i$ in the Shannon expansion of Eq. 5.6 can be set to constant values. However, this ATPG interpretation of Shannon's expansion is quite useful in the more general case of Eq. 5.7, i.e., when we expand in terms of some arbitrary function $f(\mathbf{x})$. In the general case, it is a priori *not* known if and what signals in the cofactors can be set to constant values. This however can be determined by means of a test generator.

The following notation is used. Let $y$ be an arbitrary node in a combinational network $C$ and $f$ be some Boolean function represented as combinational network. The variables of $f$ may or may not be nodes of the combinational network $C$. A new combinational network $C'$ is constructed as follows. We duplicate all nodes in the combinational network that are in the transitive fanin of $y$ so that there are two implementations of node $y$. This has been illustrated in Figure 5.13a and Figure 5.13b. One version is ANDed with $f$ the other version is ANDed with $\bar{f}$ and the outputs of the AND gates are combined by an OR gate whose output replaces the node $y$ in the original network. In the following, if $y$ and $f$ are functions of a combinational network then we refer to this construction simply by the equation $y = f \cdot y + \bar{f} \cdot y$.

Let $y$ and $f$ be Boolean functions represented in a combinational network. We propose to expand function $y$ in terms of function $f$ by the following method.

1) Network transformation: $y = f \cdot y + \bar{f} \cdot y$. (Eq. 5.9)
2) Redundancy elimination with appropriate fault list.

Note that this procedure is a special instance of the two-step procedure described at the beginning of Section 5.6. It is also one out of many possibilities to perform a Boolean division or orthonormal expansion in a combinational network. We will now investigate network transformations that are theoretically possible. The following theorem proves that by using the construction of Eq. 5.9 and redundancy elimination as the only means of transforming a network, theoretically no generality is lost with respect to manipulating a network. It turns out that the above ATPG-based expansion can be used to perform arbitrary manipulations of a combinational network.

**Theorem 5.5:** Let $y^i$ be a node of a combinational network $C^i$. The gates in this combinational network have no more than two inputs. Further, let $f^i$ be a divisor which is represented as a combinational network and realizes a Boolean function of *no more than two* variables that may or may not be nodes in $C^i$ such that

1) The transformation of node $y^i$ into $y^{i+1}$ given by

$$y^{i+1} = f^i \cdot y^i + \overline{f^i} \cdot y^i$$

and followed by

2) Redundancy removal (with appropriate fault list)

generates a combinational network $C^{i+1}$. For an arbitrary pair of equivalent combinational networks $C$ and $C'$ there exists a sequence of combinational networks $C^1, C^2, \ldots C^k$ such that $C^1 \equiv C$ and $C^k \equiv C'$ (see Appendix for a proof).

Suppose $C$ is the given combinational network and $C'$ is the combinational network that is optimal with respect to the given cost function. Theorem 5.5 states that there always exists a sequence of the specified expansion operations such that the optimal combinational network is obtained. However, it does not say which divisors shall be used when applying Eq. 5.9. As stated in the theorem, if the network has gates with no more than two inputs, it is sufficient to only consider two-input divisors created as a function of two nodes in the network. This reduces the number of divisors that (theoretically) have to be examined. However, this restriction does not imply that more complex divisors are of no use in the presented expansion scheme. If more complex divisors are used, the network is transformed in bigger steps. Theorem 5.5 does not put any restriction on the choice of divisors to transform the network. Further degrees of freedom for the expansions lie within redundancy elimination. The result of redundancy elimination depends on what faults are targeted and in which order they are processed.

Theorem 5.5 represents the theoretical basis of a *general ATPG-based* framework to logic optimization. As mentioned, redundancy elimination and the transformation of Eq. 5.9 *per se* do not represent an optimization technique. However, they provide the basic toolkit to modify a combinational network. In order to obtain good optimization results efficient heuristics have to be developed to decide what divisors to choose and how to set up the fault list for redundancy elimination. This will be described in the following.

Consider again the example of Figure 5.13. For the above expansion, the circuit transformation of Eq. 5.7 requires that all combinational circuitry in the transitive fanin of $y$ is duplicated before redundancy elimination is applied. This seems impractical and in the following, we consider special cases of the expansion where only one cofactor has to be considered. These special cases correspond to divisor functions $f$ which follow from $y$ by *implication*. For the following lemmas, let $f$ and $y$ be nodes of the combinational network $C$ such that $f$ is not in the transitive fanout of $y$. (The node $f$ must not be in the transitive fanout of $y$ in order to ensure that the circuit remains combinational after the transformation).

**Lemma 5.1:**
Consider a function: $y' = y|_1 + \overline{f}$.
Then, $y' = y$ if and only if the implication $y = 0 \Rightarrow f = 1$ is true (see Appendix for a proof).

**Lemma 5.2:**
Consider a function: $y' = f + y|_0$.
Then, $y' = y$ if and only if the implication $y = 0 \Rightarrow f = 0$ is true.

**Lemma 5.3:**
Consider a function: $y' = f \cdot y|_1$.
Then, $y' = y$ if and only if the implication $y = 1 \Rightarrow f = 1$ is true.

**Lemma 5.4:**
Consider a function: $y' = \overline{f} \cdot y|_0$.
Then, $y' = y$ if and only if the implication $y = 1 \Rightarrow f = 0$ is true.

The lemmas state that implications determine exactly those functions $f$ such that the function $y$ has only *one* cofactor with respect to $f$. In other words, in a combinational network the expansion of Theorem 5.5 can be simplified in the presence of the specified implications so that no circuitry has to be duplicated. This makes it attractive to use *implicants* as divisors for the orthonormal expansion of Eq. 5.7. Section 4.3 has introduced the notion of an implicant for multi-level combinational networks. Implicant-based transformations in multi-level networks are considered in the following theorem.

**Theorem 5.6:** Let $y^i$ be a node of a combinational network $C^i$. The gates in the combinational network can have no more than two inputs. Further, let $f^i$ be an implicant according to Definition 4.3 such that

1) the transformation of node $y^i$ into $y^{i+1}$ given by

$y^{i+1} = y^i + \overline{f}^i$ for $y = 0 \Rightarrow f = 1$
$y^{i+1} = f^i + y^i$ for $y = 0 \Rightarrow f = 0$

$$y^{i+1} = f^i \cdot y^i \qquad \text{for } y = 1 \Rightarrow f = 1$$
$$y^{i+1} = \bar{f}^i \cdot y^i \qquad \text{for } y = 1 \Rightarrow f = 0$$

followed by

2) redundancy removal (with appropriate fault list)

generates a combinational network $C^{i+1}$. For an arbitrary pair of equivalent combinational networks $C$ and $C'$ there exists a sequence of combinational networks $C^1$, $C^2$,... $C^k$ such that $C^1 \equiv C$ and $C^k \equiv C'$ (see Appendix for a proof).

Theorem 5.6 delivers the legitimacy for extending the notion of an implicant from two-level to multi-level circuits as described in Chapter 4. Transformations based on the extended notion of an implicant as in Section 4.3 allow us to perform arbitrary manipulations of a combinational network. Any transformations otherwise described by the notions of functional decomposition, kerneling, division, transduction, etc., can also be described using this extended notion of an implicant in a multi-level network. This provides a unified view for optimization of two-level and multi-level circuits.

Note that the transformations described by Lemmas 5.1 through 5.4 are closely related to the transformations in the global flow approach given by Theorem 5.3 of Section 5.4. However, an important difference between the approach presented here and the global flow is that global flow uses a more restricted set of implications than can be obtained by AND/OR reasoning techniques and that the don't cares for the cofactors in Lemmas 5.1 - 5.4 are used in a different way. In global flow there is a special construction to transform and minimize the circuit presented in Section 5.4. Here, redundancy elimination is employed. Further differences between the two approaches will be discussed in Section 5.6.2.

Lemmas 5.1 through 5.4 cover those cases where the implementation of a node $y$ in a combinational network can be replaced by a different implementation of the node with an equivalent function. As already explained in Section 4.3.2 a function at node $y$ can also be replaced by some non-equivalent function $y'$ if this does not change the function $C(\boldsymbol{x})$: $B_2^n \rightarrow B_2^m$ of the combinational network as a whole. Also mentioned in Section 4.3.2, such functions are called *permissible functions* [Muro89]. By considering permissible functions rather than only equivalent functions as candidates for substitution at each node, additional degrees of freedom are exploited as given by *observability don't cares*, defined in Section 5.3.4.

**Definition 5.10:** For an arbitrary node y in a combinational network $C$ assume the single fault $y$ stuck-at-$V$, $V \in \{0, 1\}$: If $f = U$, $U \in \{0, 1\}$, is a value assignment at node $f$ that is *necessary* to detect the fault on at least one primary output of C, then $f = U$ follows from $y = \overline{V}$ by "D-implication" and is denoted: $y = \overline{V} \xrightarrow{D} f = U$.

Let $f = 0$ be a necessary assignment for the detection of the fault $y$ stuck-at-1. With the terminology of Section 4.3.2 this can also be expressed by stating that $f$ is a (single-literal) *permissible 1-implicant* for $y$. Performing D-implications means to identify permissible implicants for some node in the network.

The conventional implications are a special case of D-implications. Replacing the implications in Lemma 5.1 through 5.4 by D-implications we obtain the following generalization:

**Theorem 5.7:** Let $f$ and $y$ be arbitrary nodes in a combinational network $C$ where $f$ is not in the transitive fanout of $y$ and both stuck-at faults at node $y$ are testable.

The function $y'$: $B_2^n \rightarrow B_2$ , $B_2 = \{0, 1\}$ with $y' = y|_1 + \bar{f}$ is a *permissible function* at node $y$ if and only if the D-implication $y = 0 \xrightarrow{D} f = 1$ is true (see Appendix for a proof).

**Theorem 5.8 - 5.10:** analogous to Lemmas 5.2 - 5.4.

Theorems 5.7 through 5.10 represent the basis for circuit transformations in our optimization method. The circuit is transformed at nodes $y$ by performing expansions in terms of functions $f$ which satisfy the implications as given in Theorem 5.6 using the extension to D-implications as given by Theorems 5.7 through 5.10.

*Recursive learning*, described in Chapter 3, permits to determine all value assignments necessary to detect a single stuck-at fault, i.e., it is a technique to perform D-implications. It can therefore be used to identify candidate functions for the above expansions. Note, however, that recursive learning can only identify those candidate functions that are already present as nodes in the network. This restriction can be overcome by using AND/OR reasoning graphs of Section 4.3 to identify general multi-literal implicants.

There usually exists a large number of implicants and permissible implicants for a node in the network and the question arises what implicants are good candidates for circuit transformations. AND/OR reasoning techniques do not only provide sets of candidate functions for which the expansions of Theorems 5.7 through 5.10 are valid, they also offer heuristic guidance. "Good" implications pointing out "good" functions can be recognized by examining the *topology* of the AND/OR graphs. This will be described in Section 5.6.2.

### 5.6.2 Heuristics to Select Implicants

This section describes heuristics to identify implications and implicants representing promising candidates for circuit transformation. First, only single-literal implicants are considered. They can be derived by recursive learning. Then, this heuristic approach will be generalized to multi-literal implicants derived from AND/OR reasoning graphs.

*A) Single-Literal Implicants:*

The following method of identifying single-literal divisors has been motivated by an observation first mentioned in [RaCo90]. *Indirect* implications indicate sub-optimality in the circuit. This is illustrated in Figure 5.14.

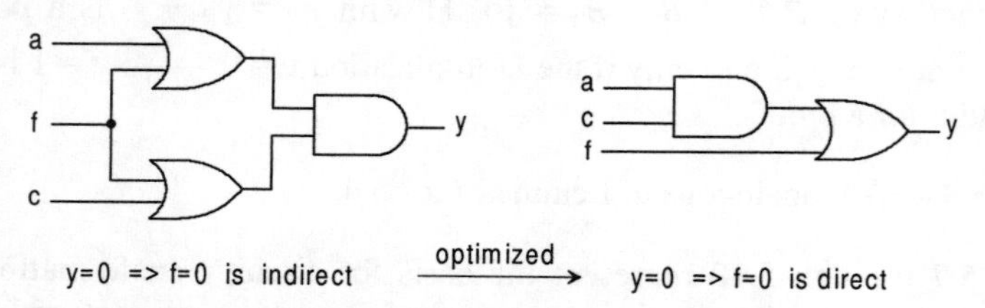


**Figure 5.14:** Indirect implication and optimization

In the left circuit of Figure 5.14, consider the initial situation of value assignments $y = 0$ for which $f = 0$ can be implied *indirectly*. This can be accomplished by means of recursive learning. Note that the existence of the indirect implication $y = 0 \Rightarrow f = 0$ is due to the fact that the circuit is not properly optimized. In the optimized right circuit which is functionally equivalent to the left circuit we note that the implication $y = 0 \Rightarrow f = 0$ is direct. One may verify that all examples of indirect implications shown in previous sections are also due to poorly optimized circuitry. Apparently, indirect implications are the key to identifying and optimizing sub-optimal circuitry. Intuitively, more the recursion needed to identify an indirect implication, better it is to attempt a logic minimization based on the expansions in Theorem 5.7.

The relationship between the complexity of deriving implications and the minimality of a combinational network represents an interesting possibility to guiding logic minimization techniques. In the following we refer to a *D-implication* $y = V \xrightarrow{D} f = U,\ U, V \in \{0, 1\}$, as *indirect* if it can be derived neither by direct implication nor by *unique sensitization* [FuSh83] at the dominators [KiMe87] of $y$. That is, the necessary assignments obtained by routines *complete_unique_sensitization()* and *make_all_implications()* with at least one level of recursion are implied *indirectly* and provide a set of promising candidates for circuit transformations.

**Example 5.12:** "good" Boolean division

Consider Figure 5.15. By recursive learning it is possible to identify the indirect implication $y = 1 \Rightarrow f = 1$. The fact that the implication $y = 1 \Rightarrow f = 1$ is *indirect* means that it is promising to attempt a Boolean division at node $y$ using the divisor $f$. This could be performed by any traditional method of Boolean division. Instead, we use the ATPG-based expansion introduced in Section 5.6.1.

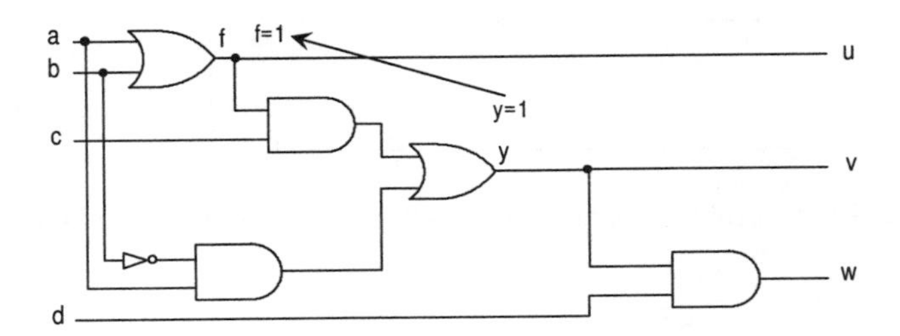


**Figure 5.15:** Combinational network C with *indirect* implication

Applying Theorem 5.7 we obtain the combinational network shown in Figure 5.16. Actually, in this case we could also apply Lemma 5.3 since $y = 1 \Rightarrow f = 1$ is obtained without using any requirements for fault propagation. Theorem 5.7. states that $y' = f \cdot y|_1$ is a permissible function for $y$. In this case, $y$ and $y'$ are equivalent. By transformation as shown in Figure 5.16 we introduce the node $y' = f \cdot y$. Since $y$ is used as a cover for $y|_1$ it is likely that the internal don't cares result in untestable single stuck-at faults. This is used in the next step.

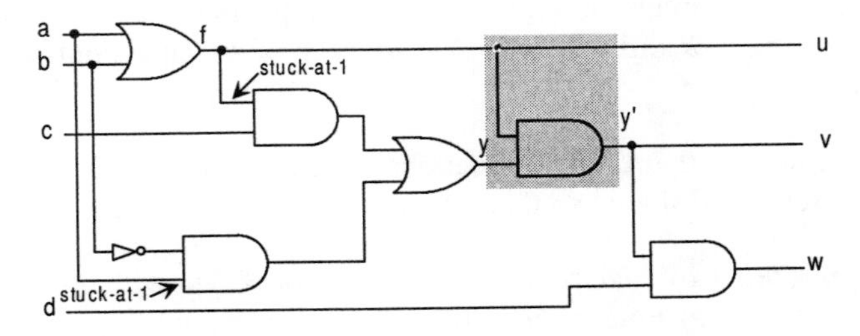


**Figure 5.16:** Circuit after transformation $y' = f{\cdot}y$ using node $f$ as divisor

By ATPG the untestable faults indicated in Figure 5.16 can be identified. Performing redundancy removal results in the minimized combinational network as shown in Figure 5.17. Note that we have to exclude from the fault list the stuck-at faults at the added circuitry in the shaded area of Figure 5.16. If we performed redundancy elimination on line $f$ in Figure 5.16 we would return to the original network.

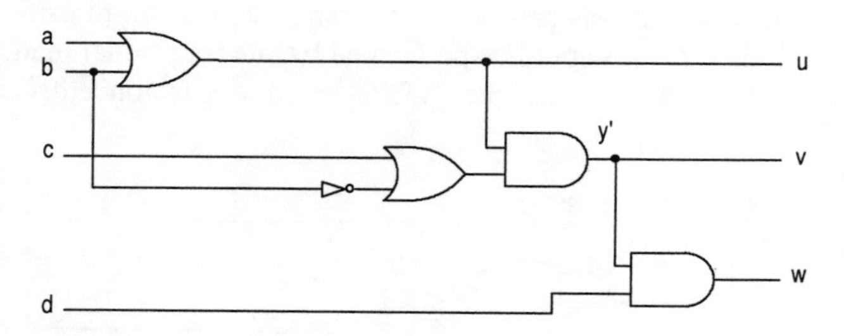


**Figure 5.17:** Combinational network after reduction by redundancy removal

The node $y$ in Figure 5.15 is implemented by $y = c(a + b) + a\bar{b}$. By *indirect* implication we identified the Boolean divisor $f = a + b$ as "promising" and performed the (non-unique) division $(c(a + b) + a\bar{b}) / (a + b)$, resulting in $y' = (a + b)\cdot(c + \bar{b})$ of Figure 5.17. Note that this is a *Boolean* - as opposed to *algebraic* - division. As the example shows, *indirect* implications help to identify good divisors that justify the effort to attempt a Boolean division.

**Example 5.13:** "common kernel extraction"

Consider the circuit of Figure 5.18. It implements two Boolean functions: $u = b(a + c) + cd$ and $v = b(f + e) + ed$, each of which cannot be optimized any further.

Note, however, that the two functions have a common cube, $b + d$, which can be extracted and shared so that a smaller circuit is obtained:

$$u = b(a + c) + cd \;=\; ab + c(b + d) \;=\; ab + cg, \text{ with } g = b + d$$
$$v = b(f + e) + ed \;=\; bf + e(b + d) \;=\; bf + eg, \text{ with } g = b + d$$

It is interesting to examine how the sub-optimality of the original circuit is reflected by the indirectness of implications.

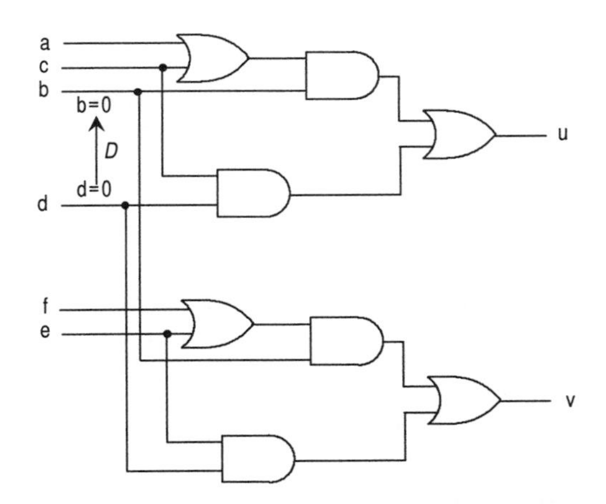


**Figure 5.18:** Extraction of common cube, $b + d$, by D-implication

Consider Figure 5.18. By recursive learning it is possible to identify the D-implication $d = 0 \xrightarrow{D} b = 0$. This means $b = 0$ is necessary for detection of $d$ stuck-at-1. As can be noted the necessary assignment $b = 0$ is not "obvious". It can neither be derived by direct implications nor by sensitization at the dominators of $d$. The reader may verify that $b = 0$ can be obtained by the learning case of recursive learning using *complete_unique_sensitization()*.

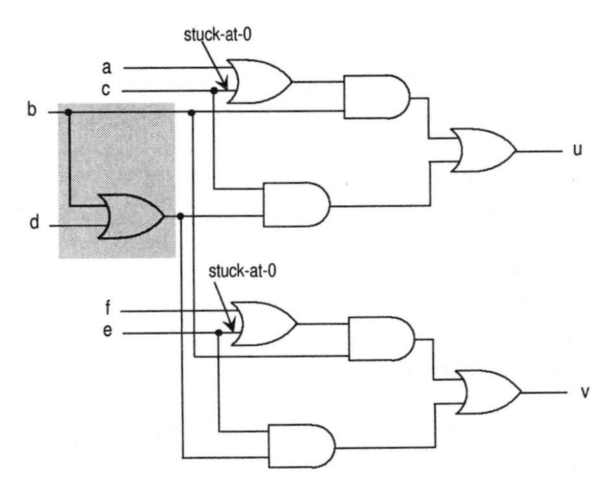


**Figure 5.19:** Replacing $d$ by permissible function $d' = b + d$

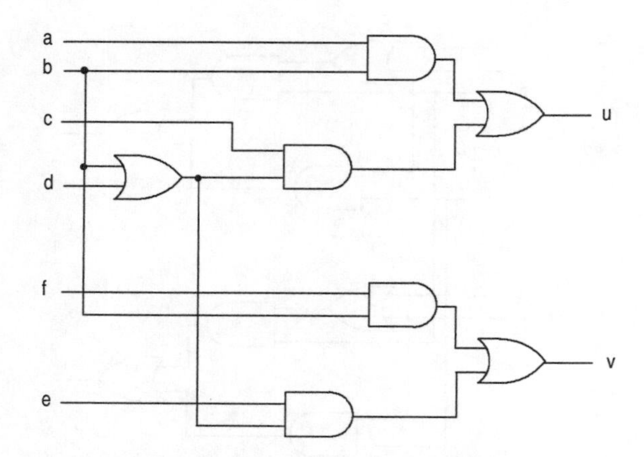


**Figure 5.20:** Optimized circuit by sharing of logic

Now the expansion is performed in the usual way. According to Theorem 5.5 the circuit can be modified as shown in Figure 5.19 and redundancy elimination yields the optimized circuit in Figure 5.20. Note that this transformation cannot be obtained by the methods of [EnCh93], [ChMa94].

As presented in Section 3.2 recursive learning consists of two techniques, *make_all_implications()* and *complete_unique_sensitization()*. It can be observed that implications obtained by *make_all_implications()* result in expansions that, in conventional terms, are most adequately described by Boolean division or Boolean resubstitution. This was illustrated in Example 5.12. If the implication is obtained by *complete_unique_sensitization()* as in Example 5.13, the expansion often performs what is commonly referred to as common kernel extraction.

A comparison with the global flow approach of Section 5.4 is illuminating. Both methods use implications for circuit transformation but there are some important differences. The method presented here selects a node $y$ in the network and then determines implicants $f$ for this node. Transformations are performed by adding the implicants $f$ to the cover of $y$. This is like in standard two-level minimization procedures as discussed earlier. By way of contrast, in global flow some signal $f$ is considered as an implicant for other nodes $y$ of the circuit and then this implicant is added to the cover of *all* nodes. The advantage of the global flow approach is that it is known after this specific construction what signals can be removed from the circuit. No test generation is needed. On the other hand, opportunities for circuit optimization can be missed. In Example 5.12 the global flow approach would consider the transformation as shown in Figure 5.16. However, it will not find all signals that become redundant. Of the two redundancies shown in Figure 5.16 global flow would only remove one, namely the one in the fanout branch of signal $f$. Therefore, one can expect that the global flow approach will perform faster than the optimization techniques presented here, however, optimization results are generally worse. Furthermore, global flow does not exploit any observability don't cares. For example, the circuit transformation of Example 5.13 cannot be obtained from global flow optimization.

Above examples also show a limitation of using implications (single-literal implicants). By implication analysis we only consider divisors that are already present as nodes in the network. Therefore, we do not completely utilize the generality of our basic approach as presented in Section 5.6.1. In the following it is shown how AND/OR reasoning graphs can be applied more generally to *create* divisors that are not present in the network.

### *B) Multi-Literal Implicants*

AND/OR reasoning trees, in principle, can generate all permissible prime implicants. Hence, they obtain any permissible function for some node in the network expressed in terms of arbitrary internal variables of the multi-level network. However, there may be a large number of prime implicants for a given node in the network, especially, if the implicants are expressed in terms of arbitrary (internal) nodes. Therefore, this sub-section is dedicated to demonstrate how the *topology* of the AND/OR trees can be used to generate certain implicants that are particularly promising for optimization.

In fact, the indirectness of implications proves to be an excellent heuristic guidance of the optimization process. This heuristic approach is now generalized. Remember that an indirect implication corresponds to an IST with *several* leaves, all belonging to the *same* value assignment. Intuitively, a subtree of the AND/OR tree which has *several* leaves that all correspond to the *same* value assignment indicates sub-optimal circuitry. If such a subtree is an IST, then we have an implication and transformations like those in Section 5.6.1 can be performed.

The motivation for the method here comes from the observation that there can be subtrees with several identical leaves (intuitively, indicating sub-optimal circuitry) which do not represent implications by themselves. However, several such subtrees together can form an IST and represent a “good“ implicant. In this section we propose a method that assembles implicants for nodes in a combinational network where the variables of the implicants correspond to the leaves of maximal subtrees with identical leaves. Such implicants can then be used for logic optimization in the same way as shown in Section 5.6.1 for indirect implications.

As proved in Section 4.3 permissible prime implicants correspond to MISTs of the D-AND/OR enumeration tree. According to the above argument, trees with many identical leaves correspond to particularly promising implicants for circuit optimization. Hence, we intend to identify a MIST with many identical leaves. To keep memory requirements low we avoid building the trees and extract the implicants only by enumeration. Therefore, in the following we show how a MIST with a maximum number of identical leaves can be extracted from the D-AND/OR enumeration tree, solely by "monitoring" the enumeration process. In Figure 4.11 of Section 4.3, the bold lines indicate a promising MIST with four leaves that only belong to two different value assignments. Therefore, the permissible implicant *bc* is promising for inclusion in a permissible function at node *a*.

It is now demonstrated how this permissible implicant for node *a* is constructed from the AND/OR tree in Figure 4.11. To avoid storing the graph we perform repeated AND/OR enumeration. In each pass we extract subtrees from the AND/OR tree that have several leaves corresponding to *one* value assignment. These subtrees correspond to promising literals included in an implicant.

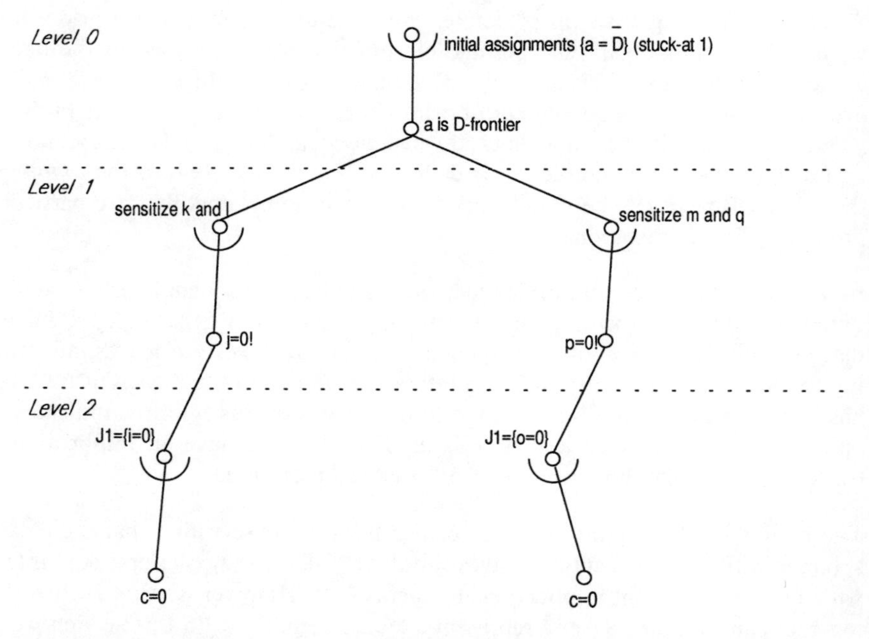


**Figure 5.21:** Literal subtree of AND/OR tree in Figure 4.11 suggests to include $c = 0$ in implicant

Figure 5.21 shows such a subtree for the assignment $c = 0$. Subtrees with identical leaves belonging to literals of an implicant will be called *literal subtrees* (*LST*) and are defined as follows:

**Definition 5.11:** A literal subtree (LST) for a variable assignment $f = V$, $V \in \{0, 1\}$, of a MIST $T$ is a subtree of $T$ such that

i) it has the same root node as $T$ and

ii) it contains all leaves of $T$ with $f = V$ and contains no other leaves.

**Example 4.4 (contd.):** The subtree shown in Figure 5.21 is an LST for literal $c$. The fact that this subtree has more than just one leaf means that $c$ is an "important contribution" to (the observable part of) the function at node $a$ and that it should be included in the implicant. An implicant is assembled step by step by identifying LSTs of large size. In the above example we can proceed as follows: We pick $c$ as the first "seed" literal in the implicant. In order to capture the part of function not covered by $c$, we now assign $c = 1$ in the circuit and re-enumerate the AND/OR tree. The resulting AND/OR tree is shown in Figure 5.22

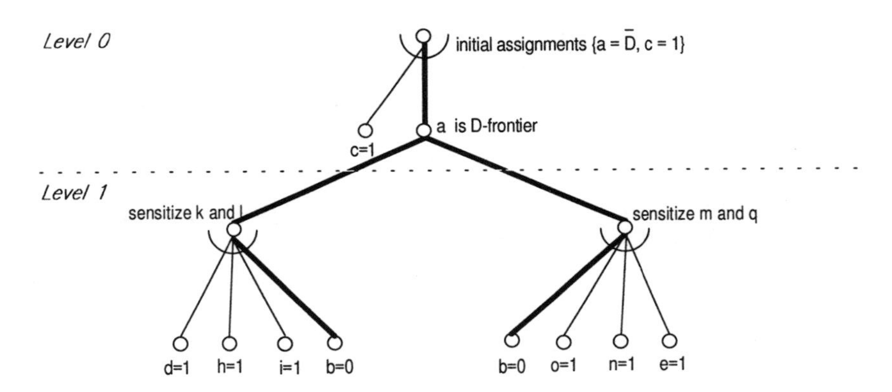


**Figure 5.22:** Subtree after assigning $c = 1$ indicates that $b = 0$ should be included in implicant

The resulting AND/OR tree suggests that variable $b$ should be included in the implicant. The subtree for variable $b$ is shown by bold lines in Figure 5.22 and represents an LST with two leaves. Other LSTs of the tree in Figure 5.22 would only have one leaf. Note that now $b = 0$ is an indirect implication derived from the new initial set of value assignments, because this time the subtree contains *all* children of the non-terminal OR nodes (there is only one in this example) that are also present in the full tree. In other words, this LST is also a MIST. Once a subtree represents an implication, the construction is finished and the implicant is complete. We obtain the $\bar{D}$-implicant $bc$ for node $a$. Consider Table 5.5 and Table 5.6 that show the pseudo-code for calculating implicants as illustrated in the above example.

```
/*  this procedure operates on a global data structure representing the circuit,
    y is a node in the circuit for which implicants are determined, r_max is the
    maximum recursion depth for AND/OR enumeration */
find_implicant(y, r_max)
{
    /* LST_candidate_list, f.LST.leaf_count.V_r, f.MIST_leaf_count.V_r ,
       f.V.mark_list, are global variables and are determined
       in and_or_based_variable_selection() */

    assign in the circuit y := 0;
    mark all gates for all levels;
    consistent = AND/OR-based_variable_selection(0, r_max);

    if (consistent = INCONSISTENT)
        return FALSE;                          /* y, stuck-at-1 is redundant */

    Select a (f = V) from the LST_candidate_list with
        maximal f.LST_leaf_count.V_r=0;        /* maximal LST size */
    if (several candidates have the maximal LST size)
        select from those one with minimal f.MIST_leaf_count.V_r=0
                                               /* minimal MIST size */
    if (V = 0) then x_1 := f else x_1 := f̄ ;

    implicant := x_1;              /* "seed" literal for forming implicant */

    i := 1;
    loop
    {
        assign in the circuit x_i :=  V̄_i;
        i := i+1;
        for (all elements (g, r) in x_i.V_i.mark_list)
            mark gate g for level r;
        consistent := and_or_based_variable_selection(0, r_max);

        if (consistent = INCONSISTENT)
            break;                             /* implicant is complete */

        Select a (f = V) from the LST_candidate_list as above;
        if (V = 0) then x_i := f else x_i := f̄ ;
        implicant := implicant · x_i;          /* add literal to implicant */
    }
    return (implicant);
}
```

**Table 5.5:** Routine to calculate implicants in a multi-level circuit

```
and_or_based_variable_selection(S, r, r_max) {
  make all direct implications for S in circuit
  and set up a list U^r of unjustified gates in event list E(S);
  if (value assignments are logically inconsistent)
      return INCONSISTENT;
  else {
      for (every signal f during the direct implication process)
          if (signal f is assigned value V, V∈{0,1})
              f.LST_leaf_count.V_r := f.LST_leaf_count.V_r + 1;
  }
  if (r<r_max) {
      for (each unjustified gate g in U^r
           which is marked in level r) { /* marked in routine find_implicant() */
          set up list of justifications gC^r;
          /* try justifications */
          for (each justification J_i ∈ gC^r)
              consistent_i := and_or_based_variable_selection(J_i, r+1, r_max );
          /* check logic consistency */
          if (consistent_i = INCONSISTENT for all i)
              return (INCONSISTENT);
          else {
            /* determine subtree sizes of LSTs and MISTs by counting leaves */
              n := number of consistent justifications;
              for (every signal f touched during implications in level r+1) {
                /* check heuristic criterion to select LST and MIST */
                  if ( f.LST_leaf_count.V_r+1 > f.LST_leaf_count.V_r or
                      (f.LST_leaf_count.V_r+1 = f.LST_leaf_count.V_r and
                       n + f.MIST_leaf_count.V_r+1 - 1 < f.MIST_leaf_count.V_r )){
                       /* take this LST and MIST */
                      f.LST_leaf_count.V_r := f.LST_leaf_count.V_r+1;
                      f.MIST_leaf_count.V_r := n + f.MIST_leaf_count.V_r+1 - 1;
                      /* mark the considered portion of the AND/OR tree */
                      f.V.mark_list := { (u, l) | unjustified gate u belongs to an
                      OR node of the selected MIST in recursion level l};
                  }
                  f.LST_leaf_count.V_r+1 := 0;
                  f.MIST_leaf_count.V_r+1 := 0;
                  if (r=0)
                      LST_candidate_list := LST_candidate_list ∪ {f=V};
              }
          }
      }
  }
  return CONSISTENT;
}
```

**Table 5.6:** Selecting variables for implicants (AND/OR enumeration shaded gray)

Table 5.5 describes routine *find_implicant()*. For simplicity, we assume that AND/OR enumeration is performed without any consideration to observability. The procedure starts with the assignment $y = 0$ (If we choose $y = 1$, then a dual technique can be formulated generating sum terms instead of product terms). Then, the AND/OR tree is enumerated as given in Table 5.6 to find large LSTs as illustrated in the example. The signals with their assignments are stored in a list (*LST_candidate_list*) and are ordered according to the size of the subtrees. The variable with the largest LST is chosen as the first variable for the implicant to be created.

If there are several variables with LSTs of the same size then we pick the one which belongs to the smallest MIST. The reason for this heuristic is that we intend to generate implicants with as few literals as possible, because they will require the least area in the circuit. If the MIST belonging to a LST is very large, we may have to add many more literals to obtain an implicant and therefore we prefer LSTs that are "almost" as large as the MIST they belong to.

Then, the loop in Table 5.5 adds more variables to the product term until the product term is an implicant. Each variable added to the product term is selected by the evaluation of the AND/OR tree in which the previously identified variables are assigned their opposite values. The loop terminates if the current variable and its assignment obtained by *and_or_based_variable_selection()* represent an implication for the previous situation of value assignments. This can either be identified during AND/OR enumeration by the same checks as in recursive learning, or as shown in Table 5.5 where the loop terminates in the next iteration due to a logic inconsistency.

Table 5.6 shows how variables are selected by AND/OR enumeration. For better readability, parts of Table 5.6 are shaded. They are identical to the AND/OR enumeration of Table 4.1. The idea is to extract variables by "monitoring" the AND/OR enumeration procedure. The non-shaded parts of Table 5.6 show operations that collect certain information during the enumeration process. The collected information essentially consists of the sizes of the LSTs and MISTs, and is gathered by counting the value assignments obtained during the consistent justifications. Further, by marking the unjustified gates for each recursion level as shown in Table 5.6, it is guaranteed that the next enumeration pass, i.e., the re-enumeration after having made a decision for a certain literal in *find_implicant()*, only considers a MIST that contains the corresponding LST. Also, note that the repeated enumeration is accelerated drastically by excluding all unjustified gates from consideration that have not been marked by the previous AND/OR enumeration pass.

**Example 4.4 (contd.):** Reconsider the circuit of Figure 4.10. The function of the circuit is given by following Boolean expressions:

$$l = ad + b(cd + f) = (a + bc)d + bf$$
$$q = ae + b(ce + g) = (a + bc)e + bg$$

By manipulating the equations, it can be noted that there exists a common kernel, $a + bc$.

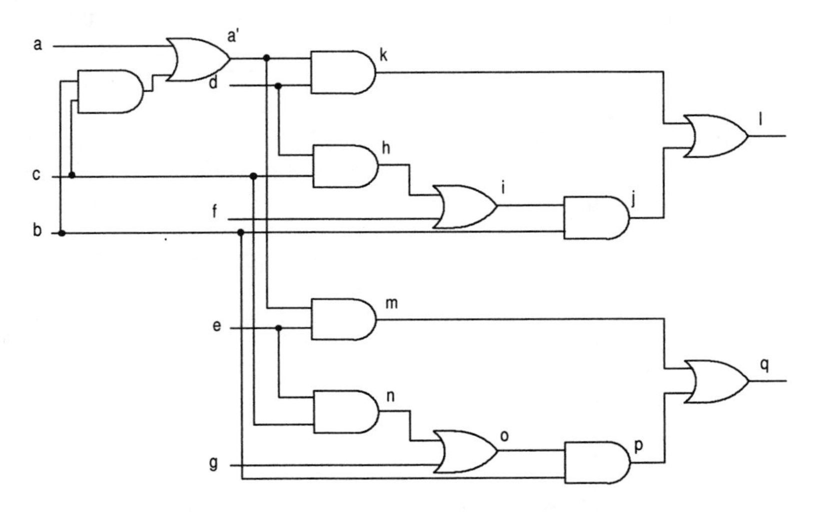


**Figure 5.23:** Adding permissible implicant *bc* at signal *a*

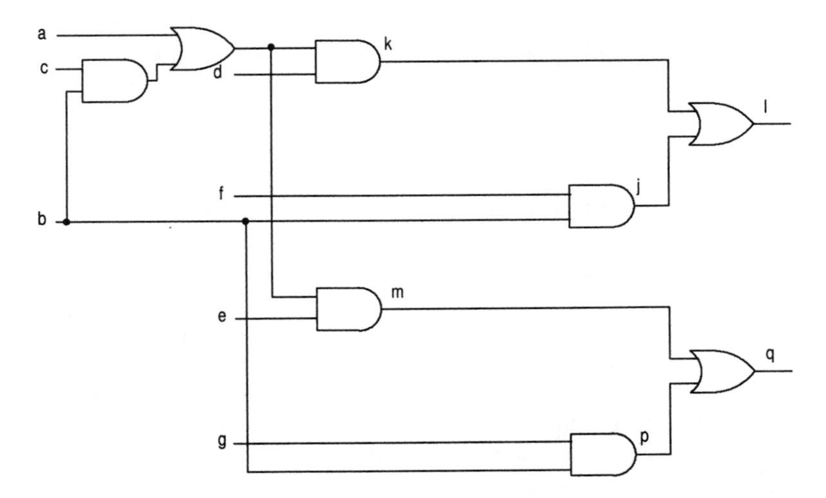


**Figure 5.24:** Circuit after redundancy removal

Minimization can be achieved by sharing this kernel. With the routines of Table 5.5 and Table 5.6 we obtain the permissible 1-implicant *bc* as was illustrated above. Note that the sub-optimality of the original circuit is reflected by the existence of a MIST with several leaves belonging to the same value assignments, $b = 0$ and $c = 0$. In this case it points out a common kernel, $a + bc$. The fact that *bc* is a permissible 1-implicant for node *a* means that the function *bc* can be inserted via an OR gate at node *a*, so that *a'* represents a permissible function in the network. This is shown in Figure 5.23. The circuit optimization procedure is as described in Section 5.6.1. If *bc* is a permissible implicant for *a* then the circuit is modified as shown in Figure 5.23. After adding the implicant to the circuit, redundancy elimination is used to reduce the circuit. This results in the circuit of Figure 5.24.

As already explained, the optimization in this example can also be obtained by an algebraic kernel extraction technique [Bray87], [RaVa90]. Note however, that the procedures based on AND/OR trees and redundancy elimination are capable of performing *general Boolean manipulations* and are not restricted to algebraic transformations.

### 5.6.3 Optimization Procedure

Table 5.7 summarizes a procedure for circuit optimization. It is a refinement of the general two-step methodology described at the beginning of Section 5.6. Logic optimization is performed by applying the described concepts to all nodes in the combinational network. The procedure moves from node to node. Experiments have shown that the optimization results are only moderately sensitive to the order in which the different circuit nodes are picked, however, best results are generally obtained by picking the nodes according to their topological level moving from primary inputs towards primary outputs. For a selected node the concepts of AND/OR enumeration are used to derive promising implicants. The candidates found promising are stored in lists and tried one after the other.

For each implicant the circuit is transformed according to the rules given in Section 5.6.1. After each transformation redundancy elimination is employed. Redundancy elimination is quite a time consuming process. Some speed-up can be gained by not considering all faults in the circuit for redundancy elimination but to restrict the search for untestable faults to those areas of the circuit where they are most likely to occur. Experimental results have shown that untestable faults occur almost always at those lines that are involved in the reasoning process of deriving the implications used for the circuit transformations.

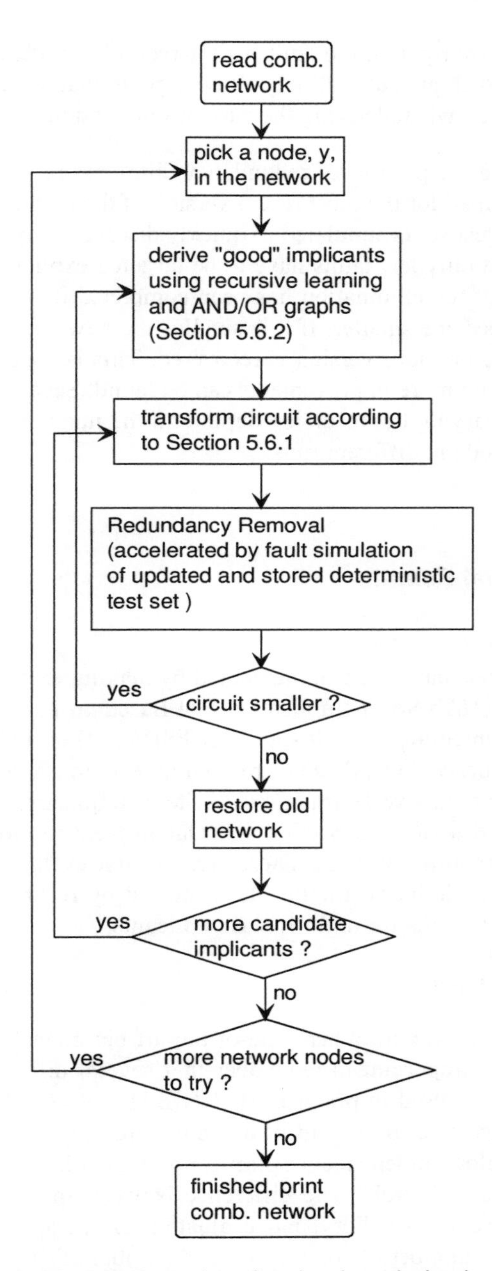


**Table 5.7:** Procedure for circuit optimization

Thus, after each transformation the fault list for redundancy identification is set up by including both stuck-at faults at only those signals that were "touched" by the AND/OR enumeration when deriving the current implication.

To further accelerate the process of redundancy elimination, the deterministic test set is always maintained for the most recent version of the circuit. After each circuit transformation this test set is simulated to quickly discard many faults from further consideration so that only few faults have to be targeted explicitly by deterministic ATPG. After redundancy elimination has been completed it is checked whether or not the circuit has become smaller. If it is smaller the new current circuit is maintained, otherwise the previous version is recovered. This is continued for all nodes in the network until no more improvements can be found. Several runs can be made through the circuit varying the recursion depth and the number of candidate implicants tried at each node in different runs.

### 5.6.4 Experimental Results

The described methods have been implemented by making extensions to the HANNIBAL tool system (HANNover Implication tool Based on Learning). For efficient fault simulation we integrated a fault simulator FSIM [24] into HANNIBAL. Logic transformations are derived by AND/OR reasoning techniques. Single literal implicants are derived by recursive learning, multi-literal implicants are determined by AND/OR graphs as described in Section 4.3. Our implementation for multi-literal implicants is at a preliminary stage. Therefore, we first evaluate our optimization tool by only using single-literal implicants identified by recursive learning. Then we show preliminary results for multi-literal implicants.

*A) Single-Literal Implicants*

We compare HANNIBAL with other state-of the art optimization tools. For a fair comparison it is very important to remember that several different ways of measuring the area costs are used in practice. HANNIBAL and RAMBO [EnCh93] operate on a gate netlist description and measure the area in terms of the number of *connections*. Technology-independent optimization tools like SIS measure the area in terms of numbers of literals. The difference between these two measures has been explained in Section 5.1. For a fair evaluation of our tool we present the results in terms of both, number of connections and number of literals.

| Name | Original | SIS_1.2 | RAMBO | HANNIBAL | |
|---|---|---|---|---|---|
| | #conn. (#lit.) | #conn. (#lit.) | #conn. (#lit.) | #conn. (#lit.) | CPU h : min : s |
| c1355 | 992 (562) | 778 (554) | 837 (546) | 764 (540) | 0:01: 31 |
| c1908 | 1059 (769) | 708 (535) | 784 (551) | 696 (511) | 00:04:38 |
| c2670 | 1559 (1023) | 1082 (752) | 1520 (816) | 1064 (701) | 00:03:37 |
| c3540 | 2226 (1658) | 1649 (1288) | 1810 (1331) | 1628 (1221) | 01:13:57 |
| c432 | 296 (270) | 247 (205) | 271 (207) | 207 (181) | 00:00:55 |
| c499 | 368 (562) | 776 (554) | 837 (546) | 348 (540) | 00:00:16 |
| c5315 | 3492 (2425) | 2548 (1731) | 3201 (1851) | 2661 (1779) | 00:15:39 |
| c6288 | 4768 (3313) | 4695 (3337) | 3834 (3294) | 3723 (3252) | 00:29:18 |
| c7552 | 4734 (3087) | 3457 (2312) | 3385 (2188) | 2542 (1826) | 00:36:00 |
| c880 | 640 (433) | 594 (413) | 643 (410) | 578 (400) | 00:00:44 |
| 9symm | 387 (237) | 384 (186) | 324 (217) | 286 (217) | 00:03:19 |
| alu2 | 669 (453) | 507 (361) | 509 (359) | 347 (274) | 00:05:29 |
| alu4 | 1299 (855) | 975 (694) | 1006 (722) | 826 (646) | 00:36:40 |
| apex6 | 1214 (835) | 1074 (743) | 1327 (759) | 1000 (697) | 00:03:54 |
| apex7 | 410 (289) | 331 (245) | 412 (251) | 309 (229) | 00:00:24 |
| dalu | 3533 (2610) | 1364 (979) | 2007 (1344) | 1710 (1102) | 02:10:18 |
| frg2 | 2244 (2005) | 1182 (887) | 1734 (1157) | 1315 (982) | 00:27:35 |
| pair | 2795 (1803) | 2356 (1602) | 2594 (1636) | 2155 (1636) | 00:15:29 |
| rot | 1085 (764) | 928 (672) | 1093 (662) | 834 (633) | 00:02:20 |
| term1 | 773 (456) | 235 (170) | 363 (248) | 208 (149) | 00:00:44 |
| ttt2 | 434 (324) | 303 (219) | 300 (191) | 204 (148) | 00:00:26 |
| x1 | 627 (445) | 409 (298) | 503 (333) | 392 (298) | 00:00:58 |
| x3 | 1589 (1133) | 1101 (787) | 1547 (985) | 1110 (1035) | 00:03:59 |
| x4 | 843 (1607) | 512 (380) | 777 (449) | 583 (400) | 00:03:34 |

**Table 5.8:** Logic minimization results for HANNIBAL (Sun SPARC 5)

| Name | SIS_1.2 (script.rugged) | SIS_1.2 (script.rugged) + HANNIBAL | |
|---|---|---|---|
| | #conn. (#lit.) | #conn. (#lit.) | CPU time h : min : s |
| c1355 | 778 (554) | 759 (543) | 00:01: 20 |
| c1908 | 708 (535) | 690 (516) | 00:02:38 |
| c2670 | 1082 (752) | 1021 (773) | 00:02:37 |
| c3540 | 1649 (1288) | 1571 (1144) | 00:34:18 |
| c432 | 247 (205) | 212 (165) | 00:00:28 |
| c499 | 776 (554) | 763 (540) | 00:01:26 |
| c5315 | 2548 (1731) | 2425 (1679) | 00:07:16 |
| c6288 | 4695 (3337) | 3720 (3210) | 00:31:28 |
| c7552 | 3457 (2312) | 2516 (1778) | 00:22:24 |
| c880 | 594 (413) | 589 (416) | 00:00:48 |
| 9symm | 384 (186) | 227 (178) | 00:01:49 |
| alu2 | 507 (361) | 355 (279) | 00:04:44 |
| alu4 | 975 (694) | 776 (596) | 00:26:38 |
| apex6 | 1074 (743) | 999 (687) | 00:02:36 |
| apex7 | 331 (245) | 294 (224) | 00:00:14 |
| dalu | 1364 (979) | 1171 (735) | 00:44:02 |
| frg2 | 1182 (887) | 1052 (834) | 00:05:46 |
| pair | 2356 (1602) | 2149 (1509) | 00:15:35 |
| rot | 928 (672) | 822 (641) | 00:01:45 |
| term1 | 235 (170) | 183 (131) | 00:00:10 |
| ttt2 | 303 (219) | 225 (165) | 00:00:24 |
| vda | 688 (615) | 630 (566) | 00:09:36 |
| x1 | 409 (298) | 377 (287) | 00:00:31 |
| x3 | 1101 (787) | 995 (758) | 00:03:48 |
| x4 | 512 (380) | 482 (357) | 00:00:47 |

**Table 5.9:** Results for HANNIBAL after pre-processing with SIS (Sun SPARC 5)

For RAMBO and HANNIBAL the number of literals (factored form) has been obtained by reading the optimized circuits into SIS and post-processing them such that a technology-independent factored form is obtained. For this purpose we used a SIS script obtained from [Chen94] which performs some standard network manipulations. To count connections for SIS we map the optimized circuit to a generic library which contains the basic gates that are allowed in our netlist description. Note that comparing connections or literals may slightly bias the results. Since RAMBO and HANNIBAL optimize in terms of connections whereas SIS uses literals, comparing connections can bias the results in favor of HANNIBAL and RAMBO. Comparing literals gives a certain advantage to SIS. Therefore, for all circuits we always present both area measures.

In all experiments HANNIBAL passes through the circuit four times performing expansions at every node where recursive learning can identify indirect implications. The recursion depth is '1' for the first two passes and '2' for the final two passes. We also experimented with higher depth of recursion. It turned out that recursion depth higher than '2' did not lead to improved optimization results because the same transformation which can be derived by high recursion depth can usually also be obtained by a sequence of local transformations derived by small recursion depth. Also, for larger designs a recursion depth of '4' and higher is usually not affordable in terms of CPU-time.

Table 5.8 shows results for SIS_1.2, RAMBO_C and HANNIBAL. SIS_1.2 is run using *script.rugged* which includes the powerful techniques of [RaVa90] and [SaBr91]. No pre-optimization is used to process the circuits in RAMBO and HANNIBAL. As can be noted, for most benchmark circuits HANNIBAL produces smallest circuits. This is quite remarkable because it shows that most circuit manipulations performed by conventional technology-independent multi-level minimization techniques are covered by the netlist transformations presented in Section 5.6.1 using single-literal implicants. In particular, heuristic guidance by indirect implications proved remarkably powerful.

In the next experiment it is examined how much optimization is possible by HANNIBAL if the circuits are pre-processed by SIS. As shown in Table 5.9 substantial area gains are possible in many cases. For 7 out of 25 circuits the gain is more than 20%. Also note that the CPU-times for HANNIBAL are significantly shorter in many cases if the circuits are first run through a technology-independent minimization.

Finally, we also compare our results with [ChMa94]. In [ChMa94] the circuit is mapped to a library with only 2-input gates and results are only shown after preprocessing with SIS. Table 5.10 shows the results for RAMBO (taken from [ChMa94]), PERTURB/SIMPLIFY [ChMa94] and HANNIBAL when the area was measured in terms of 2-input gates. We take a subset of the above benchmarks for which results are shown in [ChMa94] and all circuits are pre-optimized by SIS. As can be noted HANNIBAL obtains circuits that are smaller than or equal to those obtained by RAMBO or PERTURB/SIMPLIFY. Note that the results of HANNIBAL and RAMBO could be further improved over [ChMa94] if the cost function was changed to optimize the number of 2-input gates.

| Name | SIS + | | |
|---|---|---|---|
| | RAMBO | PER./SIM. | HANNIBAL |
| | # 2-input gates | # 2-input gates | # 2-input gates |
| c3540 | 988 | 938 | 922 |
| c5315 | 1458 | 1321 | 1288 |
| c6288 | 2334 | 1883 | 1872 |
| c7552 | 1761 | 1426 | 1292 |
| alu2 | 366 | 281 | 157 |
| alu4 | 700 | 555 | 426 |
| apex6 | 647 | 543 | 525 |
| rot | 569 | 452 | 452 |
| term1 | 203 | 113 | 103 |
| ttt2 | 174 | 118 | 117 |
| x3 | 617 | 552 | 542 |

**Table 5.10:** Experimental comparison with [ChMa94]

Further experiments confirmed the heuristic that *indirect* implications indicate promising divisors. We examined how many indirect D-implications existed in the circuits before and after optimization. For the ISCAS85 circuits, Table 5.11 shows the number of indirect D-implications that have been identified by recursive learning with depth ‘2’ for the original circuits, the circuits optimized by SIS and the circuits optimized by HANNIBAL. We note that HANNIBAL reduces the number of indirect D-implications drastically for all circuits. Interestingly, in most cases this is also true for SIS, which is based on an entirely different approach to optimization. This indicates that optimization in general is related to reducing the number of *indirect* implications. The results of Table 5.11 reflect that many (but not all) “good” divisors for optimization can be obtained by *indirect* implication. Also note that Table 5.11 explains why the CPU-times for HANNIBAL are generally shorter if HANNIBAL is run after SIS. If SIS is used first there are fewer indirect implications and hence less expansions need to be performed.

| Circuit | Original | SIS (script.rugged) | HANNIBAL |
|---|---|---|---|
| c432 | 568 | 308 | 134 |
| c880 | 580 | 269 | 138 |
| c1355 | 942 | 320 | 72 |
| c1908 | 6,239 | 755 | 835 |
| c2670 | 4,041 | 617 | 419 |
| c3540 | 16,770 | 18,082 | 10,767 |
| c5315 | 9,383 | 3,149 | 3,157 |
| c6288 | 3,516 | 3,275 | 618 |
| c7552 | 27,644 | 3,189 | 1,118 |

**Table 5.11:** Indirect D-implications before and after optimization

*B) Multi-Literal Implicants*

The AND/OR graph based techniques to identify multi-literal implicants have been applied to the problem of PLA factorization. This application is interesting because this task cannot be accomplished in a satisfactory way by only using single-literal implicants.

Table 5.12 shows preliminary results for some two-level MCNC benchmark circuits that are factorized into a multi-level description. Since our implementation does not accept external don't cares, we only selected examples that are completely specified. The results of HANNIBAL are compared with SIS1.2 (using *resub -a, simplify -d* followed by *script.rugged*). The area is measured in terms of number of connections based on a generic library of the basic gate types. For both tools we show results for the same fixed settings (single run of *script.rugged* for SIS) and interactive use (column "multi run" for SIS and column "best" for HANNIBAL). Column "RL" shows the results if only single literal implicants (recursive learning) are used.

CPU-times (*h:min:sec*) are preliminary. As described in Section 5.6.2, implicants are determined only by repeated enumeration without actually building the graphs. If the graphs are actually constructed (at the cost of memory) implicants can be determined much faster by simple operations on the graph. Future implementations may investigate appropriate trade-offs between memory and time as was outlined in Section 4.4. An advantage of the presented AND/OR trees is that they need not be constructed to their full size in order to be useful. In these experiments, AND/OR trees have been examined only up to a recursion depth of '3'. This however proved sufficient to obtain the shown optimization results.

| PLA factorization | | SIS1.2 (script rugged) | | HANNIBAL | | | |
|---|---|---|---|---|---|---|---|
| results | | single run | multi run | fixed settings | | best | RL |
| name | # conn. | # conn. | # conn. | # conn. | CPU-time | # conn. | # conn. |
| 5xp1 | 369 | 164 | 159 | 79 | 0:01:58 | 78 | 237 |
| 9sym | 609 | 320 | 206 | 152 | 0:17:00 | 83 | 609 |
| clip | 1055 | 195 | 187 | 110 | 0:10:24 | 90 | 520 |
| con1 | 32 | 30 | 30 | 27 | 0:00:01 | 27 | 30 |
| duke2 | 995 | 540 | 510 | 416 | 1:12:33 | 355 | 612 |
| e64 | 2144 | 253 | 253 | 253 | 0:15:19 | 253 | 253 |
| misex1 | 154 | 77 | 77 | 59 | 0:00:51 | 55 | 81 |
| misex2 | 206 | 121 | 121 | 121 | 0:02:47 | 111 | 134 |
| o64 | 195 | - | - | 195 | 0:00:08 | 195 | 195 |
| rd53 | 176 | 52 | 52 | 36 | 0:00:42 | 34 | 99 |
| sao2 | 501 | 192 | 190 | 116 | 0:05:38 | 108 | 195 |
| vg2 | 914 | 124 | 124 | 115 | 0:03:22 | 112 | 141 |

**Table 5.12:** Experimental results for multi-literal implicants (Sun SPARC 5)

The experimental results confirm our conjecture that topological properties of AND/OR reasoning graphs can be used to guide an optimization process. In the examined cases HANNIBAL obtained (sometimes significantly) better optimization results than SIS1.2. This also remained true when we ran SIS interactively, repeating *script.rugged* multiple times.

Our experiments also demonstrate the practical relevance of the theoretical result in Theorem 4.1. This is illustrated by the example of the MCNC benchmark circuit *o64*. This circuit is small, nevertheless it is impossible to build an OBDD for this circuit. No literal in the circuit description appears in more than one product term (hence the SOP is unate) and all prime implicants are essential. Therefore, optimization is impossible, either with two-level or with multi-level minimization techniques. SIS1.2 runs out of memory in both *script.rugged* and *script.algebraic* after wasting about ten minutes of CPU-time in each script. However, since the circuit is unate, Theorem 4.1 applies. This explains why HANNIBAL has no problem with this example. The AND/OR reasoning trees for this circuit are of linear size and of trivial structure. No MISTs exist where several leaves belong to the same value assignment. Therefore, no promising implicants can be generated. The fact that this circuit cannot be further optimized is determined very quickly and only little CPU-time is wasted.

A main advantage of the proposed optimization approach is that it operates directly on the structural netlist description of the circuit so that the technology impact of the performed transformations can be easily evaluated, permitting a better control of the optimization process with respect to the specific goals of the designer. The reported results have only considered area minimization and it has been demonstrated that the presented method is competitive with conventional technology-independent minimization techniques. Based on this approach, tools have been presented targeting other design goals: optimization for low power consumption has been examined in [PrCh96] and optimization for random pattern testability in [ChPr95]. It has not yet been considered how the presented techniques can handle circuits with *complex gates* in an efficient way. The implementation of HANNIBAL is limited to handling only the basic gate types, AND, OR, NAND, NOR, INV, XOR. Future work should therefore extend the techniques to handle complex gates so that arbitrary libraries can be processed.

Chapter 6

# LOGIC VERIFICATION

This chapter describes a further application of the algorithms developed in Chapters 3, 4 and 5. It is motivated by the growing interest in formal verification methods and outlines new ways of approaching *logic verification*. Logic verification is the problem of deciding whether two combinational circuits implement the same Boolean functions. Throughout this chapter we ignore the possible presence of external don't care conditions, i.e., only completely specified functions are considered. Extensions to external don't care conditions are possible but not considered in this discussion.

No general verification technique is known which performs well for all classes of circuits. Therefore, we must resort to investigating special properties of practical verification problems that can be exploited to formulate effective procedures. This leads to a hybrid verification approach combining well-known *functional* representations like OBDD (Section 1.3.4) and the *structure* oriented implication-based procedures of Chapters 2, 3, 4 and 5.

## 6.1 Motivation

With increasingly complex algorithms and software involved in VLSI design, the synthesis process can never be guaranteed to be completely free of errors. Any complex software can contain bugs and even a well-debugged tool, having worked correctly for many applications, may introduce unforeseen errors in future applications. Therefore, it is crucial to verify a circuit including that obtained by automatic synthesis tools against the initial specification.

Design errors can also be introduced by the often needed interference of the human designer. Designers sometimes resort to manual modifications, or incorporate tailor-made software to fulfill the special design requirements. Changes at a later stage of the design process are required if the specification has been modified slightly, or if a finished product must be updated to satisfy special requests. Such modifications are referred to as engineering changes (ECs). Because ECs may be conducted manually, this makes them both expensive and prone to errors. The designer must, therefore, have efficient tools to check the functional correctness of designs after ECs.

The approach to be described is useful in verifying circuits (or sub-functions of a circuit) after incremental design changes or ECs as they often occur in a synthesis environment. Further, an efficient verification method is also an important integral part of synthesis if modifying and updating a given design shall become an automated process (*incremental* synthesis). Approaches to incremental synthesis and automatic design error corrections have already been reported in [FuKa91], [PoRe93], [Bran94].

Logic verification belongs to the set of most difficult problems in the field of computer-aided circuit design. Even verification of small designs can lead to enormous computation times, requiring large amounts of memory. The difficulty of the logic verification problem arises from having to explore the full functionality of both design and specification in order to formally prove their functional equivalence. Logic verification, therefore, is usually attempted by *functional* methods; i.e., methods that try to completely capture the function of a circuit so that a comparison is possible. Much progress has been made with the introduction of Binary Decision Diagrams (BDDs) [Aker78] and Ordered Binary Decision Diagrams (OBDDs) [Brya86] ( Section 1.3.4). OBDDs permit a Boolean function to be represented in a compact and canonical form. If the circuits to be compared can be represented by OBDDs, the logic verification problem is solved; the circuits are functionally equivalent iff their OBDDs are isomorphic. Construction of OBDDs, however, is not always easily accomplished. The size of the OBDDs is highly sensitive to variable ordering, and for some circuits (like multipliers), the size of the OBDDs grows exponentially with the size of the circuit.

*Structural* techniques can be considered as an alternative to functional approaches. These techniques exploit the structural information of the given circuit implementations. The AND/OR graphs of Chapter 4 do not only represent the function (or sub-functions) of a circuit, they also maintain important structural information. Since the reasoning in recursive learning is strictly guided by the topological connectivity of the gate level circuit description recursive learning and the AND/OR reasoning techniques of Chapter 4, in this sense, are referred to as structural methods.

Significant progress can be achieved in applying structural techniques to logic verification. As shown in [Bran93], [Kunz93] structural approaches to logic verification can perform extremely well, provided the circuits have some structural similarity. For circuits that are dissimilar these techniques can fail. BDD-based verification, on the other hand, is independent of the structural representation of the individual designs and hence does not rely on structural similarity. However, these methods too can fail because of exponential memory requirements. Therefore, in this chapter, a new methodology is described to effectively combine structural and functional methods for logic verification to mutually exploit the advantages of both paradigms. In particular, this hybrid approach [KuPr96], [ReKu95] is based on the structural method of [Kunz93] and a functional method using the OBDDs of [Brya86].

The suggested approach to logic verification exploits the general characteristics of a typical design process. A mathematical model describing the design process is presented in [AaSt94], shown schematically in Figure 6.1.

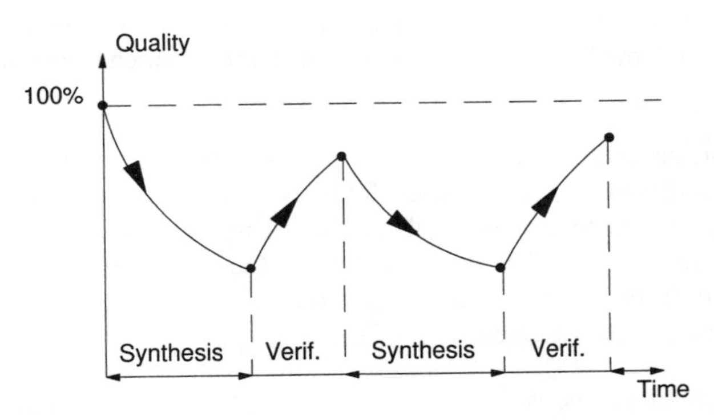


**Figure 6.1:** Typical design process (source: Aas et al., [AaSt94])

In [AaSt94], the design process is viewed as a sequence of "atomic operations" that constitute different design steps. Such design steps either can be performed manually, or automatically by synthesis tools. Each design step must be followed by an accurate and efficient verification step to achieve high design quality at low costs. It is important to note that the design process is of incremental nature as it consists of many small steps. This has important consequences for the verification steps because two successive intermediate designs, as they occur along the synthesis process, can be expected to have some degree of “similarity”. The fewer the atomic operations that are packed into a synthesis step the more similarity exists between the circuits. In particular, it can be expected that manual changes and ECs (they are particularly prone to errors) let large portions of the circuit remain unaffected.

*Similarity* refers to signal values in one circuit correlated with signal values in the other circuit if both circuits are given the same inputs. More precisely, the approach to be described makes use of the fact that values for some signals in circuit A *imply* values for signals in circuit B. As a special case, this happens when the two circuits under comparison contain internal nodes that are functionally equivalent. Intuitively, the verification problem must become much easier if there is such similarity between circuits.

## 6.2 A Hybrid Approach to Logic Verification

The previous discussion suggests a hybrid approach; i.e., a combination of structural and functional methods to be most effective. There is a wealth of structural and functional approaches that look promising to be combined for logic verification. Essentially, as proposed in [KuPr96] this chapter describes a framework of structural and functional methods and shows how different elements can be put together.

For the structural portion of this framework, it is proposed to use recursive learning. Based on structural information extracted from the circuit by recursive learning, the original circuits are divided into sub-circuits using the partitioning criterion stated below. This is coupled with a functional, OBDD-based approach to prove the equivalence of those sub-circuits. Figure 6.2 briefly outlines the general skeleton of the approach to be presented.

Let A and B be the circuits to be verified. Assume that both circuits are cut vertically so that A is split into $A_i$ and $A_o$ and B is split into $B_i$ and $B_o$. The portion indexed "i" represents the circuit partition at the primary inputs and the portion indexed "o" represents the circuit partition at the primary outputs of the original circuit.

*Criterion for circuit partitioning:* A cut through circuits A and B is *permissible* if the equivalence of circuits $A_o$ and $B_o$ implies the equivalence of circuits A and B.

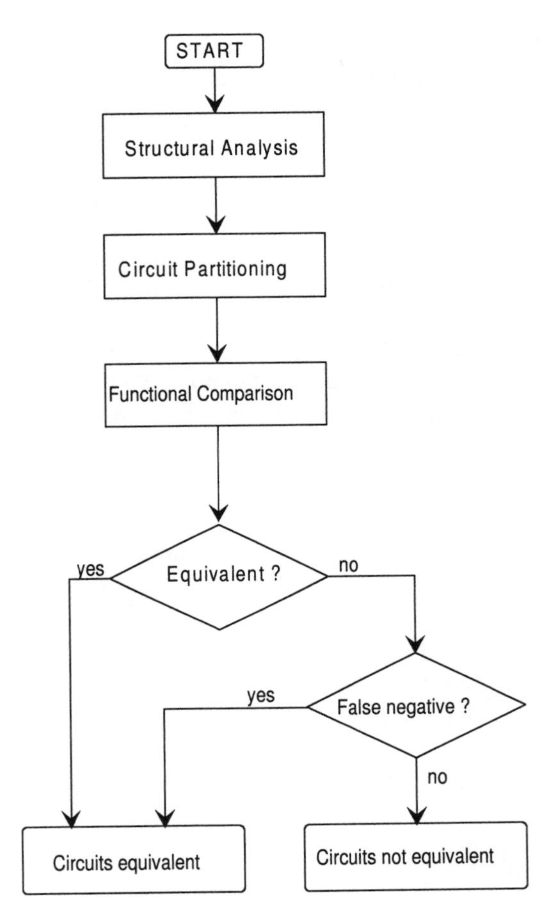


**Figure 6.2:** Hybrid structural and functional approach

One simple means of obtaining a partitioning that fulfills this criterion is to cut each circuit only through nodes which have an equivalent node in the other circuit, illustrated in Figure 6.3. It is obvious that the two circuits on the left are equivalent. Now, consider the dashed line indicating cuts in both circuits which satisfy our criterion of partitioning. This simple case demonstrates that the cut lines marked 'X' in the two circuits are equivalent. The resulting sub-circuits, $A_o$ and $B_o$, are shown on the right in Figure 6.3. Clearly, since the new pseudo-input X in circuits $A_o$ and $B_o$ is a functionally equivalent signal in the original circuits A and B, the equivalence of the circuits $A_o$ and $B_o$ implies the equivalence of the original circuits A and B and hence this partitioning fulfills the required criterion. It is important to note, however, that the opposite is not true; if circuits $A_o$ and $B_o$ are not functionally equivalent, this does not imply the non-equivalence of A and B. As can be noted in Figure 6.3, functions $y_1$' and $y_2$' are not equivalent although the original circuits are. This phenomenon shall be referred to as the *false negative* problem, following [BeTr91], and represents an unavoidable difficulty if functional and structural methods are to be combined.

Cutting through equivalent nodes is not the only way to obtain circuit partitions that fulfill the above cutting criterion. The above cutting criterion does not require that the output nodes of $A_i$ and $B_i$ are equivalent. The output functions at $A_i$ and $B_i$ may be different as long as the difference is not observable at the outputs of the original circuits A and B. Therefore, the concept of *permissible functions* [Muro89], as given in Section 4.3.2, is promising for identifying a larger set of possible cuts. This can be accomplished with the notions of ATPG [Bran93] or by recursive learning using D-implications as defined in Section 5.6.1.

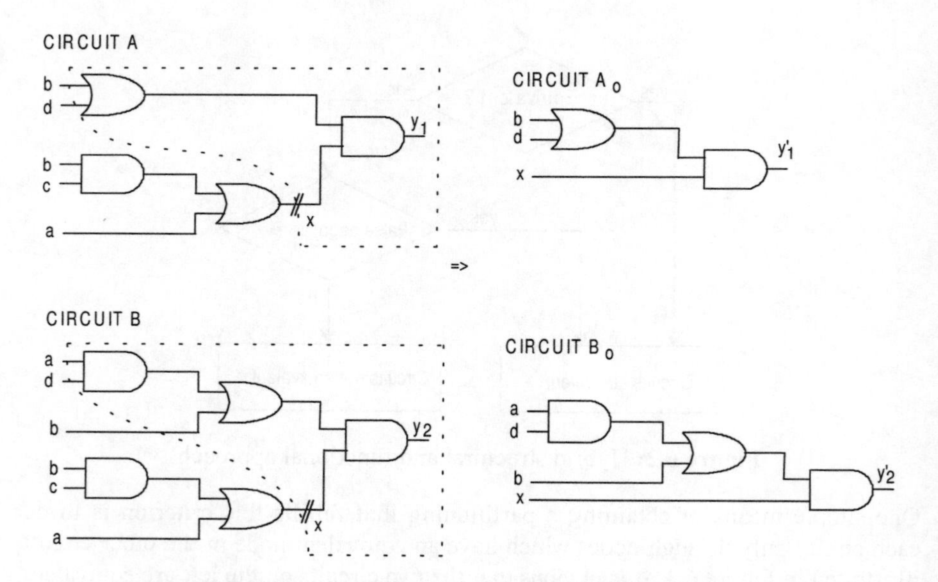


**Figure 6.3:** Circuit partitioning and false negative

Figure 6.2 summarizes the proposed approach to logic verification: first a structural analysis is performed to determine a good circuit partitioning. This is described in detail in Section 6.3. Once the circuits are cut, the smaller partitions are compared for equivalence. Experiments suggest that false negatives are not uncommon, the reason being that the created pseudo-inputs are not always independent. The BDDs so formed may contain some combinations of pseudo-inputs which are inconsistent in the original circuit; hence they represent a don't care set for the partitioned circuit. A method to tackle this problem appears in Section 6.4. Note that other functional representations like FDDs [KeSc92], free BDDs [ShDe93], OKFDDs [Drec94] or IBDDs [BiJa94], could also be used for the functional portion of the verification procedure.

In the following sections we suggest three basic ingredients that can be incorporated into a *framework for logic verification*. These are:

1) Methods to exploit structural similarity between designs (e.g., recursive learning)
2) Methods for efficient representation of Boolean functions (e.g., OBDDs)
3) Methods to solve the false negative problem arising from combining 1) and 2).

For each of these ingredients a specific technique is chosen and is incorporated into the framework. Note that significant improvement over the presented methods are still possible by adding greater sophistication in the implementation of any of the individual techniques.

## 6.3 Exploiting Structural Properties in Verification

The first ingredient necessary for effective verification procedures is the ability to exploit structural similarity. The work described here uses recursive learning to derive equivalent nodes for the purpose of partitioning the circuits as explained in Section 6.2. In principle, recursive learning may complete the verification task alone, if it can establish the functional equivalence of the primary output nodes.

The proposed verification procedure relies on combining the circuits as shown in Figure 6.4. Obviously, by this construction the logic verification problem is reduced to solving the Boolean satisfiability problem for output signal, $e$. The circuit shown in Figure 6.4 has been termed *miter* by Brand [Bran93].

In principle, the complete implication procedure of Table 3.3 represents a complete method to check the Boolean satisfiability of $e$: if the precise implications for $e = 1$ produce a conflict, then it follows that $e \equiv 0$ and the two circuits are equivalent. If no conflict occurs, the implication procedure determines all value assignments necessary for a distinguishing vector. However, what maximum depth of recursion is required in order to solve the problem? As pointed out in Section 3.2.1, the maximum depth of recursion required to identify all necessary assignments is related to the *size* of the structure related to the redundancy in the circuit. In practical circuits; i.e., circuits that realize a certain function, usually the redundant structures are rather small, so that only few recursions are needed to perform complete implications. In Figure 6.4 though there is an "artificial" circuit which does not serve any practical function. If circuits A and B are functionally equivalent, then the resulting circuit represents a very large redundancy and in this case, generally it seems intractable to perform all implications.

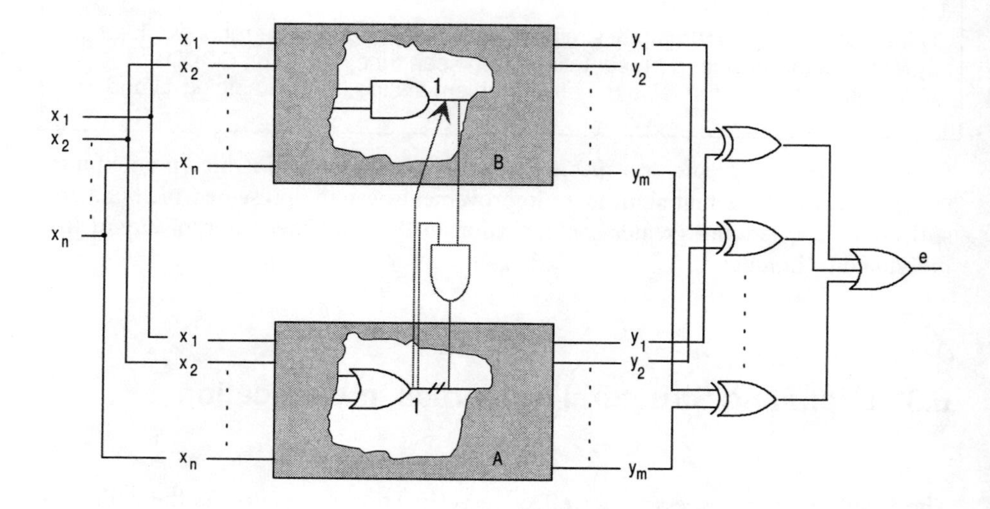


**Figure 6.4:** Logic verification by implications

This leads back to the discussion of Section 6.1. Fortunately, recursive learning can make use of a special aspect: in many cases, "similarities" exist between the two circuits under consideration. As mentioned, such similarities can be expressed as (usually indirect) implications between signals of different circuits, schematically depicted in Figure 6.4. Recursive learning can identify these implications, if they exist. Note that these implications immediately indicate the functional equivalence of internal nodes (as a special case).

In the following we discuss two ways to use implications in order to capture structural similarity in a miter. The first possibility is to simply store the implications. The second possibility is to perform optimizing transformations in the miter using implications as decribed in Section 5.6.1. The advantage of the latter approach is that it allows us to take into account observability don't cares.

### 6.3.1 Exploiting Structural Similarity by Storing of Implications

Figure 6.5 shows the flowchart of a verification procedure to exploit structural similarity by implications. At first, the miter is built as shown in Figure 6.4. Essentially, the verification process consists of two phases that can be repeatedly called. In phase 1, the procedure passes through both circuits to identify and store indirect implications. At every gate, the algorithm assigns the logic signal value which makes the gate unjustified (e.g., 0 at AND) such that more than one justification exists for that gate. Note that if 1 is assigned at the output of an AND gate, direct implications can be performed. Since our only interest is in storing indirect implications at gates where the direct implications "get stuck", 0 is assigned at the output of AND and NOR gates, and 1 is assigned at the outputs of OR and NAND gates. For XOR gates, both values are assigned one after the other. After the assignment, *make_all_implications()* is called to perform implications. If signal values are learned, i.e., if indirect implications exist, these are stored at the respective gate. This procedure is repeated for every gate in both circuits. Two aspects are very important for the efficiency of this pre-processing phase:

1) The gates $g_i$ must be selected in an appropriate order. Before some gate $g_i$ is analyzed, all gates in the transitive fanin of $g_i$ must have been treated. This ordering is necessary to make maximum use of prestored indirect implications.

2) In phase 1, *make_all_implications()* is not only used to identify and store indirect implications for later reference in phase 2, but it is crucial that *make_-all_implications()*, itself makes use of previously stored indirect implications.

After phase 1 has been completed, the actual verification algorithm begins in phase 2. In [Kunz93] a test generator for phase 2 has been used; however, phase 2 also leaves room to apply the powerful concepts of modern functional representations like OBDDs. OBDDs can be applied if the prestored indirect implications are used to derive a circuit partitioning that fulfills the criterion of Section 6.2. The implications between circuits A and B immediately yield the internal equivalencies and a circuit partitioning can be derived. (A detailed description of phase 2 appears in Section 6.4). If the verification problem is still too complex and the base algorithm for logic verification in phase 2 has to abort, phase 1 is repeated with higher depth of recursion to identify more indirect implications. This continues until either a distinguishing vector is generated, or the circuits are proven equivalent, or the search is aborted because a user-defined maximum recursion depth for the preprocessing phase is exceeded. Note that phase 1 and 2 each individually represent a complete algorithm for logic verification. However, to effectively exploit the properties of practical verification problems we need the combination of both methods.

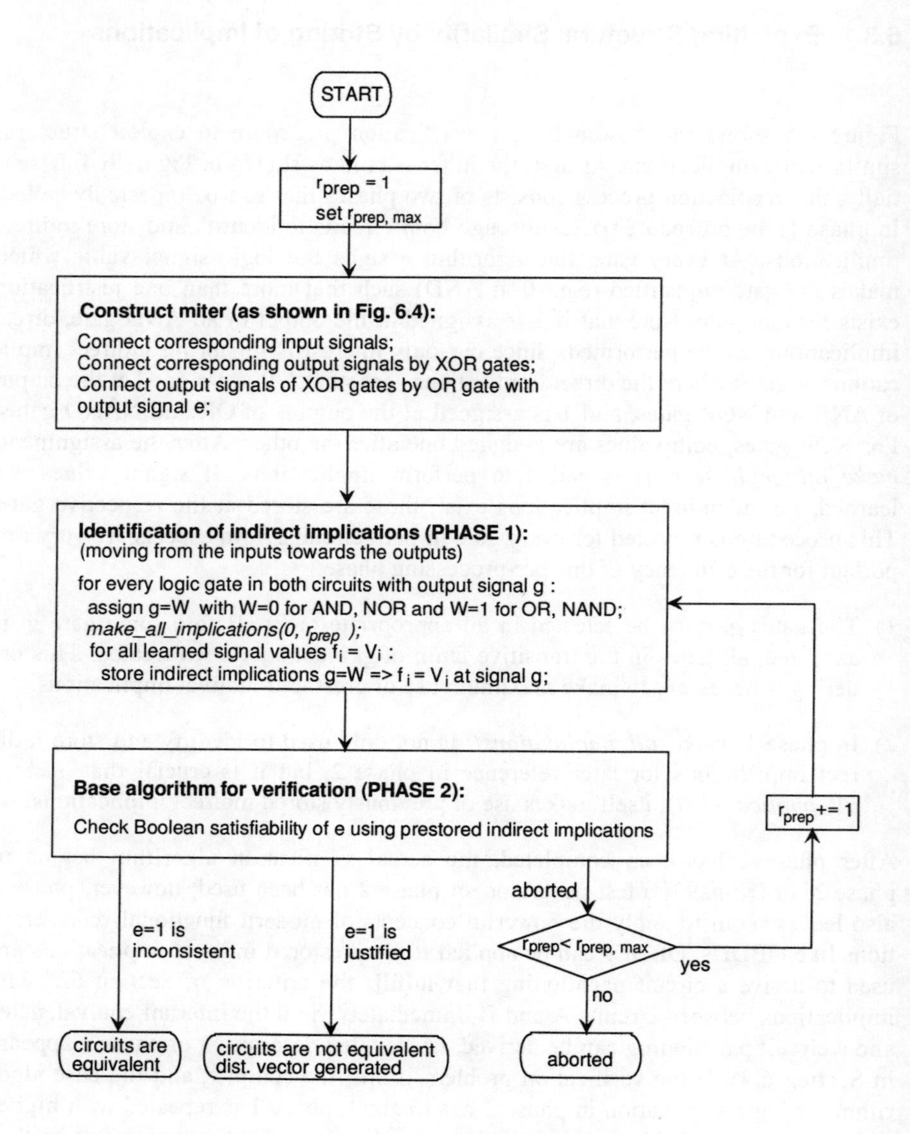


**Figure 6.5:** Flowchart of verification procedure by storing implications

**Example 6.1:** To illustrate the verification procedure an example of [BeTr89] is used, as shown in Figure 6.6. In phase 1, indirect implications are derived between the two circuits to be compared. The procedure is as follows:

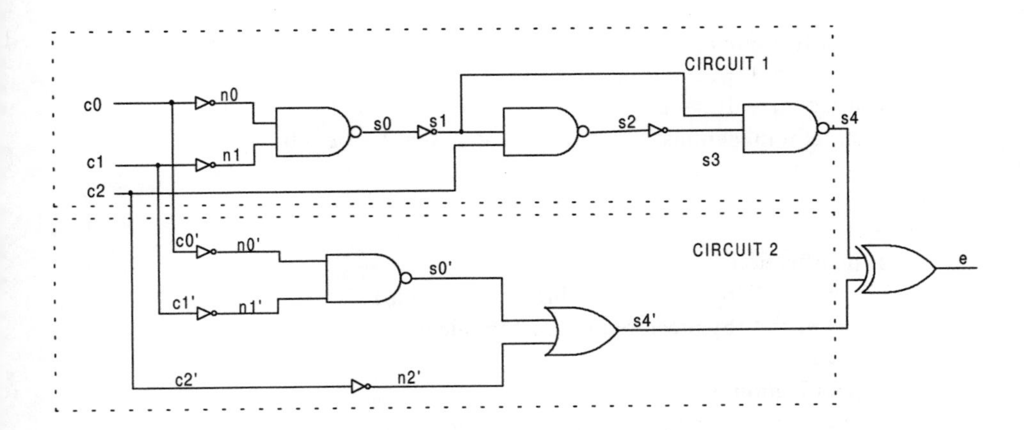


**Figure 6.6:** Example circuits [BeTr89]

*Circuit 1*: Assign the appropriate value to the output signal of every gate with two or more inputs and use *make_all_implications(0,1)* to determine indirect implications. For simplicity, internal signal values are only listed if they belong to a gate with more than one input:

$s_0 = 1: \Rightarrow s_2 = 1 \Rightarrow s_4 = 1$

1. justification: $n_0 = 0 \Rightarrow n_0' = 0$
   $\Rightarrow s_0' = 1$
   $\Rightarrow s_4' = 1$
   $\Rightarrow e = 0$
2. justification: $n_1 = 0 \Rightarrow n_1' = 0$
   $\Rightarrow s_0' = 1$
   $\Rightarrow s_4' = 1$
   $\Rightarrow e = 0$

learned: $s_0' = 1, s_4' = 1, e = 0$
indirect implications: $s_0 = 1 \Rightarrow s_0' = 1, s_0 = 1 \Rightarrow s_4' = 1, s_0 = 1 \Rightarrow e = 0$
(stored at $s_0$)

$s_2 = 1: \Rightarrow s_4 = 1$

1. justification: $s_1 = 0 \Rightarrow s_0 = 1$
   $\Rightarrow s_0' = 1$ (by prestored indirect implication)
   $\Rightarrow s_4' = 1$ (by prestored indirect implication)
   $\Rightarrow e = 0$

2. justification: $c_2 = 0 \Rightarrow n_2' = 1$
$\Rightarrow s_4' = 1, \Rightarrow e = 0$
learned: $s_4' = 1, e = 0$
indirect implications: $s_2 = 1 \Rightarrow s_4' = 1, s_2 = 1 \Rightarrow e = 0$
(stored at $s_2$)

$s_4 = 1$:
1. justification: $s_1 = 0 \Rightarrow s_0 = 1$
$\Rightarrow s_0' = 1$ (by prestored indirect implication)
$\Rightarrow s_4' = 1$ (by prestored indirect implication)
$\Rightarrow e = 0$
2. justification: $s_3 = 0 \Rightarrow s_2 = 1$
$\Rightarrow s_4' = 1$ (by prestored indirect implication)
$\Rightarrow e = 0$ (by prestored indirect implication)
learned: $s_4' = 1, e = 0$
indirect implications: $s_4 = 1 \Rightarrow s_4' = 1, s_4 = 1 \Rightarrow e = 0$
(stored at $s_4$)

*Circuit 2*: Similarly, we obtain the indirect implications:
$s_0' = 1 \Rightarrow s_0 = 1, s_0' = 1 \Rightarrow s_2 = 1$
$s_0' = 1 \Rightarrow s_4 = 1, s_0' = 1 \Rightarrow e = 1$, and
$s_4' = 1 \Rightarrow s_2 = 1, s_4' = 1 \Rightarrow s_4 = 1, s_4' = 1 \Rightarrow e = 0$

Output signal, $e = 1$:
1. justification: $s_4 = 1$ and $s_4' = 0$
$\Rightarrow e = 0$ (by prestored indirect implication), inconsistent
2. justification: $s_4 = 0$ and $s_4' = 1$
$\Rightarrow e = 0$ (by prestored indirect implication), inconsistent
learned: $e = 0$

*final result:* $e = 0$, circuits are *equivalent*

In this example, phase 1, with recursion depth 1, is sufficient to prove that the circuits are equivalent. It is straightforward to derive equivalent nodes from the prestored indirect implications. Note that only one pass is performed from the inputs towards the outputs, identifying all internal equivalences without any a priori knowledge or heuristics to identify promising candidates.

**Example 6.2:** It is interesting to observe that even in circuits with less structural similarity and no internal equivalences, it is still possible to identify logic implications between signals as illustrated in Figure 6.7. The upper circuit results from the lower circuit by a simple factorization as can frequently occur during the synthesis process. It is interesting to observe that many indirect implications exist between signals in the upper and lower circuits although the two do not contain internal equivalences and have entirely different structures.

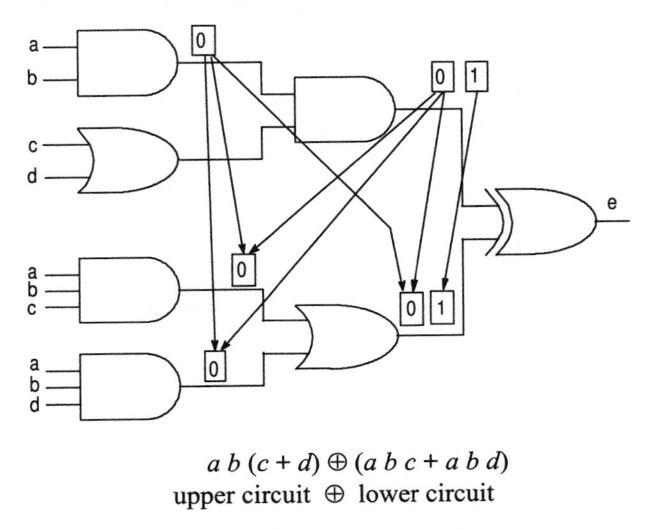


$a\,b\,(c + d) \oplus (a\,b\,c + a\,b\,d)$

upper circuit $\oplus$ lower circuit

**Figure 6.7:** Indirect implications after a typical synthesis operation

## 6.3.2 Exploiting Structural Similarity by Miter Optimization

The previous section has described that structural similarity between two designs can be exploited by identifying implications between different signals of subcircuits and by storing these implications at the respective nodes. Similarly, as proposed by Brand [Bran93], the complexity of the verification problem can also be reduced by identifying signals in one circuit which can be used to *substitute* signals in the other circuit. In [Bran93] these substitutions are found by means of ATPG. An important advantage of that technique [Bran93] is that it exploits observability don't cares, i.e., node substitutions do not always result in equivalent functions but may create *permissible* functions (see Sections 4.3.2 and 5.6.1) so that a wider spectrum of circuit transformations is considered.

Whether we make physical connections between the circuits or store implications, the two have a similar effect: they simplify the reasoning for the satisfiability solver by introducing "short cuts" between the circuits so that the satisfiability solver need not fully exhaust both circuits.

Covering a wide spectrum of structural similarity is very important, especially, if the circuits being compared do not contain a lot of similar or identical logic. The approach of [Bran93] requires that lines in one circuit can be replaced by lines in the other circuit exploiting observability don’t cares. For the method of Section 6.3.1 [Kunz93] it is required that there exist logic implications, e.g., $f = 1$ in circuit A implies $g = 1$ in circuit B. This can be a looser requirement than demanding a substitution but on the other hand this does not exploit observability don’t cares.

Taking all of this into account suggests that the verification problem should be simplified effectively by performing logical transformations in the miter (but not limited to substitutions) so that logic common to the two designs can be extracted and shared. If the circuits are equivalent then one circuit must eventually be merged into the other circuit. As a special case, the substitutions of [Bran93] perform such an operation. More generally, any known synthesis technique can be used to accomplish this task. The general goal is to *optimize* the miter. If the miter is reduced to a constant 0 the two circuits are proved equivalent. If this is not (or only partially) possible then it must be attempted to generate a distinguishing vector using ATPG.

If the circuits have a fair amount of structural *similarity* the miter can be optimized by a sequence of fairly *local* circuit transformations. If the circuits become less similar, then deriving these transformations becomes more and more complex and it becomes important to fully exploit the range and power of modern synthesis techniques. The advantage of formulating verification as a *miter optimization* problem is that the power of modern synthesis techniques becomes available to the difficult problem of logic verification.

As experimentally confirmed in Section 5.6, circuit transformations derived by indirect implications cover a large spectrum of circuit manipulations performed in standard synthesis procedures [Bray87]. Further, since implications permit an easy and effective guidance of the optimization process we can base the structural portion of our verification procedure on the optimization procedure of Section 5.6.2. This combines the advantages of the method in [Kunz93], namely that implications represent a looser requirement than node substitutions, and the method of [Bran93] that has a looser requirement by using observability don’t cares. Observability don’t cares are exploited in the implication based optimization approach of Section 5.6 by identifying *D-implications*.

Optimization of a miter has special characteristics that are discussed in the following assuming the optimization procedure of Section 5.6.2.

*Selecting implications:*

As described in Section 6.3.1, one must attempt to identify implications that are valid between two nodes that belong to *different* sub-circuits of the miter. The corresponding transformations introduce a sharing of logic between the circuits. This is schematically shown in Figure 6.4. Enforcing a sharing of logic between the circuits has two benefits: it generally reduces the size of the miter, and it tends to increase the degree of similarity in the remaining unshared parts of the circuits when the original circuits are equivalent. If the two networks are forced to share the same sub-functions that reduces the "freedom" in the implementation of the remaining parts. This is illustrated in the following example.

> **Example 6.3:** Figure 6.8 shows two circuits to be verified as equivalent. The circuits are combined to form a miter. For reasons of clarity, we depict the circuits without the extra logic of the miter. Consider signal $a$ in the upper circuit and signal $z$ in the lower circuit. By recursive learning it is possible to identify the D-implication $a = 0 \xrightarrow{D} z = 0$. The reader may verify that any test for $a$, stuck-at 1 produces the value assignment $z = 0$. According to Section 5.6.1 we can perform the circuit transformation as shown in Figure 6.9.
>
> In the transformed circuit, untestable faults are identified as indicated in Figure 6.9. Removing these redundancies leads to the circuit of Figure 6.10. Note that this transformation has not only introduced a sharing of logic between the two circuits but has reduced the overall size of the miter. It has also increased the degree of structural similarity in the remaining unshared portions of the circuit. As a matter of fact, in this example the remaining circuit portions are now structurally identical and can be shared by a sequence of very simple circuit transformations.

*Substitution:*

Often, a lot of CPU-time can be saved by restricting the circuit transformations to node substitutions. In Example 6.3, node $a$ in the upper circuit is substituted by node $z$ in the lower circuit after removing the redundancy $a$ stuck-at 0. Often it may be sufficient to restrict all transformations to only finding such substitutions [Bran93]. In general, a substitution between two nodes $y$ and $f$ can be performed by the implication-based transformations of Theorems 5.7 through 5.10, if redundancy elimination is restricted to the appropriate fault at signal $y$. This is faster than considering all faults in the circuit, but on the other hand it overlooks miter transformations which cannot be obtained by a simple substitution. Therefore, we pass through the circuit several times. In the early passes of the verification procedure the redundancy check is restricted to the node to be substituted. In later passes redundancy elimination is performed for the whole circuit.

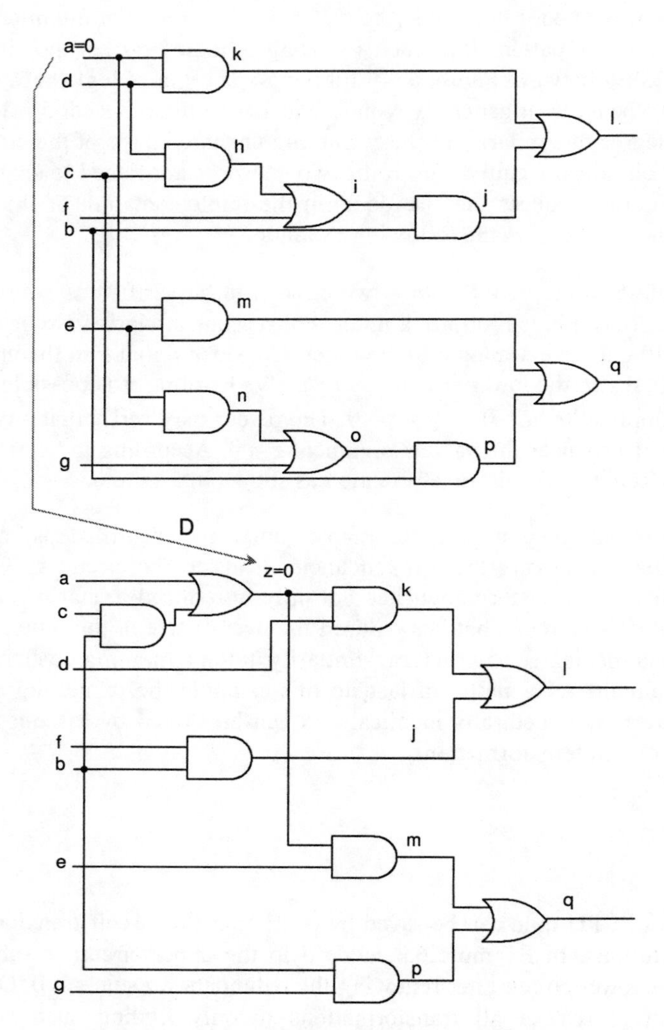


**Figure 6.8**: Circuits to be verified in Example 6.3

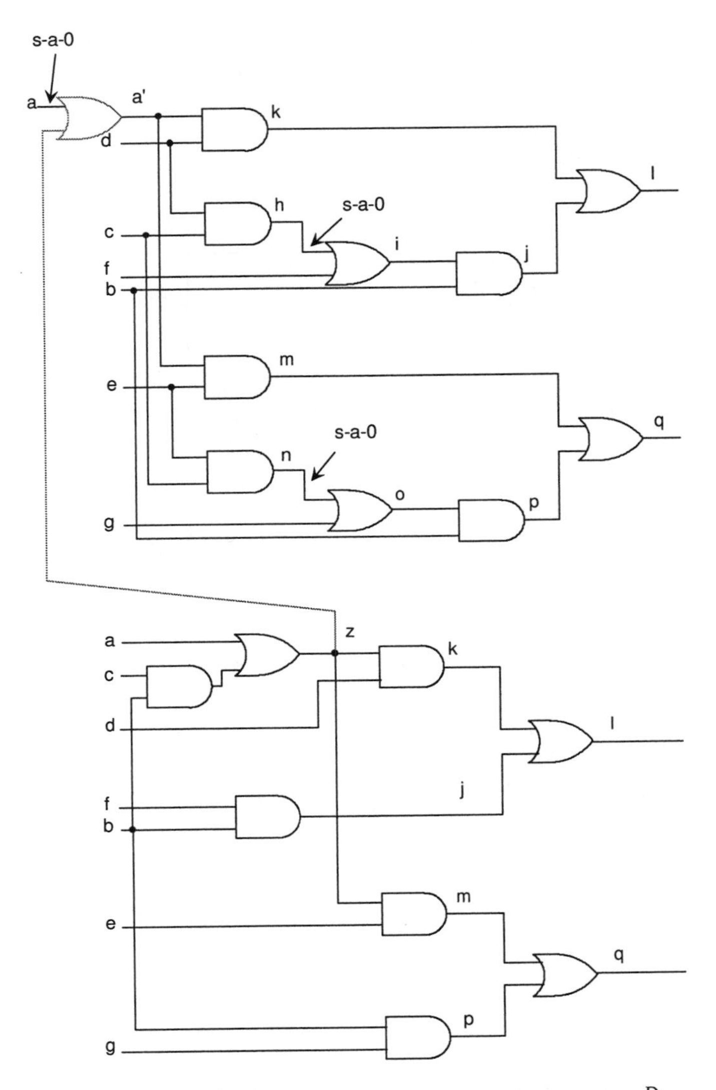


**Figure 6.9:** Circuit transformation for D-implication $a = 0 \xrightarrow{D} z = 0$

*ATPG in a miter:*

Another important aspect should be considered when running the optimization approach of Section 5.6 on a miter. The method heavily relies on evaluating circuit transformations by ATPG. However, for many target faults in the two circuits under comparison the ATPG problem becomes a lot more difficult when the circuits are connected to form a miter. In fact, a large number of faults becomes redundant, but proving these redundancies has practically the same complexity as the verification problem itself. The reason for this is the global reconvergence created by the miter. Therefore, the ATPG tool may waste a lot of time on numerous target faults which eventually have to be aborted.

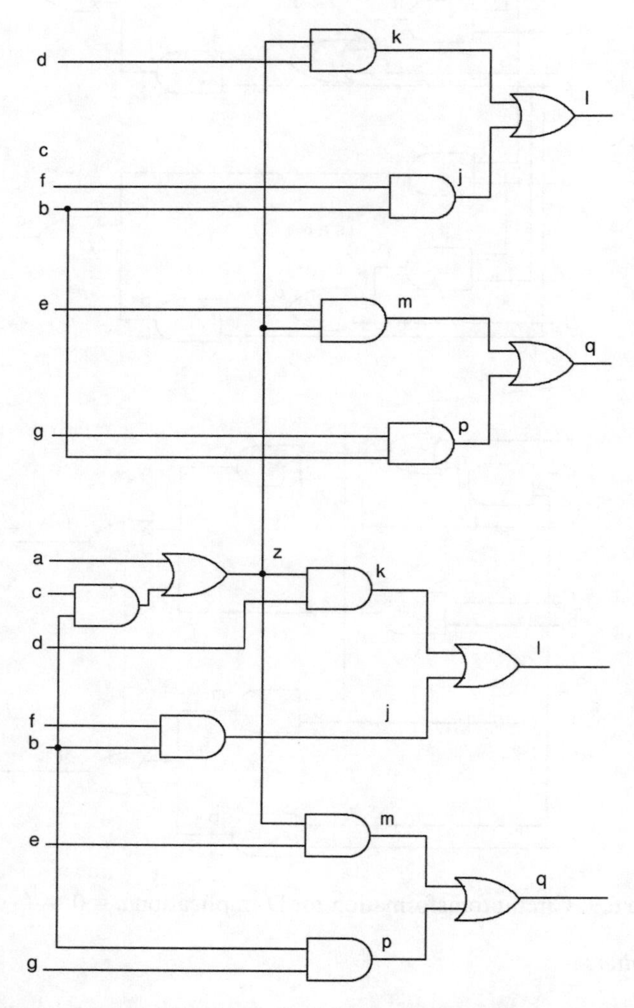


**Figure 6.10:** Miter sub-circuits after redundancy elimination

Note that we are only interested in the local redundancies introduced by the transformations of Theorems 5.7 through 5.10 but not in those redundancies caused by the global miter reconvergence. Therefore, the effect of the global miter reconvergence on the ATPG process is eliminated by the following simple trick: when performing ATPG or fault simulation faults are declared "detected" as soon as the fault signal has reached the outputs of the sub-circuits, i.e., if it has reached the inputs of the XOR-tree that forms signal *e*. In Figure 6.4 these signals are labeled $y_1$ to $y_m$. Alternatively, the XOR-portion of the miter could be removed during the ATPG-procedure. This is in fact very important for an efficient ATPG-process.

## 6.4 Functional Comparison and Solving False Negatives

The functional portion of the verification framework exploits the power of binary decision diagrams. Graph representations of Boolean functions are currently subject of extensive investigation and a variety of graph representations have been reported in recent literature. The described work uses OBDDs [Brya86]. However, other graph representations could be used in this verification methodology in a similar way.

Figure 6.11 shows the flow of the verification process. This is similar to Figure 6.2, but gives a more detailed description of phase 2. The routines *consistent_satisfy()* and *include_don't_cares()* will be described in this section.

First we partition the original circuits into smaller sub-circuits. The partitions are chosen such that the portions at the primary inputs of circuits that have already been established as equivalent by the structural phase are "cut off". A simple tracing procedure can be used for this purpose. Using a depth-first search from the outputs, the procedure traces until an equivalent node or primary input is encountered and marks it. Equivalent nodes are determined by implications if the method of Section 6.3.1 is used. If the miter is optimized as described in Section 6.3.2 the equivalent nodes correspond to the nodes shared between the two designs. All signals marked by the tracing procedure are treated as independent pseudo-inputs to the traced part of the circuit. This is performed for one circuit and the partition is mapped onto the other circuit using the corresponding equivalent nodes. In some cases this mapping may not result in a complete cut in the other circuit. In those cases the outputs whose cone does not contain a complete cut is traced in a similar fashion and the cut is made complete.

The OBDDs for the outputs are calculated for each circuit in terms of these pseudo-inputs that identify the respective circuit's partition. The OBDDs are built, using the *apply* operation [Brya86] by traversing the circuit from the pseudo-inputs towards primary outputs and building intermediate temporary OBDDs at each node's output. The OBDDs of the respective outputs are compared for equivalence. If they are isomorphic, the circuits are proved to be equivalent and if they are not isomorphic, then, a new OBDD is formed by first XOR-ing the OBDDs for the respective outputs and then forming an OR of all XOR-ed OBDDs.

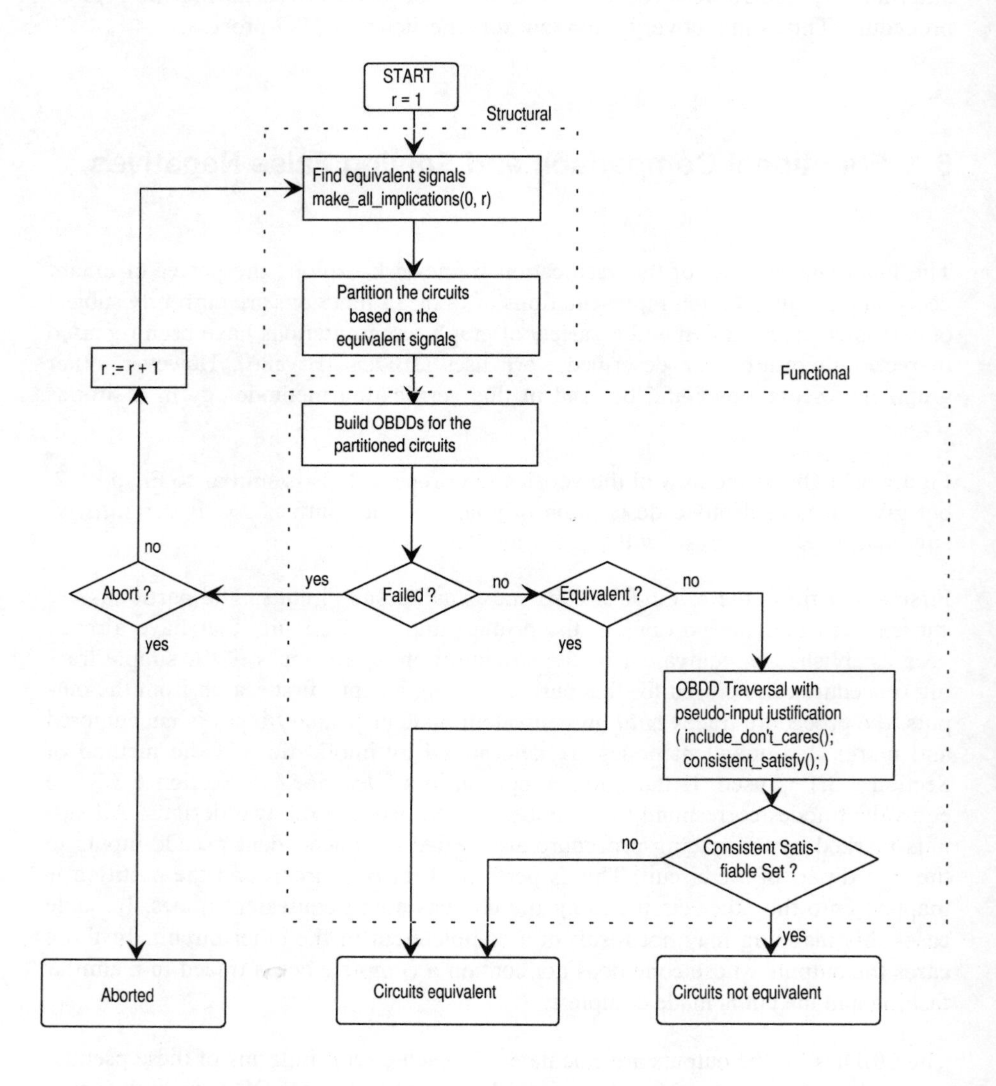


**Figure 6.11:** Block diagram of program flow in hybrid approach

### 6.4.1 Pseudo-Input Justification

The basis for the procedure to solve false negatives is the OBDD for the XOR-ed partitioned circuits (partitioned miter). Note that the size of this OBDD is quite independent of the size of the original circuits because it represents the false negative situation and not the circuits. Every path in this OBDD to the terminal node 1 corresponds to a set of value assignments at the pseudo-inputs of partitioned circuits for which the two partitioned circuits behave differently. The goal of the following procedure is to prove that such value assignments cannot occur at the internal nodes of the unpartitioned circuits and hence the corresponing path in the OBDD indicates a false negative. Of course, if these value assignments can actually occur in the unpartitioned circuits, then the two circuits are indeed non-equivalent.

> **Definition 6.1:** A *satisfying set* for the partitioned miter with a single output, *y*, refers to a set of value assignments at the inputs of the partitioned circuits, (pseudo-inputs) for which *y* evaluates to 1. If these value assignments do not produce an inconsistency in the unpartitioned circuit, it is called a *consistent satisfying set.*

The procedure is listed in Table 6.1. First, a satisfying set is found by traversing the OBDD for the partitioned miter. Then an attempt is made to find an input vector at the primary inputs of the original circuit to justify this satisfying set. If it cannot be justified, OBDD traversal is continued and a new satisfying set is found. The process is repeated until, either there is no satisfying set to be justified, which means that the circuits are equivalent, or a distinguishing vector is generated.

The recursive function *consistent_satisfy()* takes an OBDD node as an argument and finds a consistent satisfying set. This function is called with the root node of the OBDD for the partitione miter. We use the notation of [Brac90]. If a *var* is a BDD node then the BDD node pointed to by the "one" branch of *var* is represented as *var.high* and the BDD node pointed to by the "zero" branch is represented as *var.low*. First, we check the argument for a constant one or a constant zero. If it is a constant one then function *justify()* is called to justify the value assignments forming the satisfying set. If no justification exists, it is then not a consistent satisfying set and the OBDD traversal is continued. A signal value one is assigned and if its implications produce inconsistency, it is erased and signal value zero is tried. If this produces a conflict, no consistent satisfying set can be found along this path and the function returns. If the signal assignment does not produce any inconsistency, the traversal is continued by calling *consistent_satisfy* with *var.high* or *var.low* as the argument, depending on the signal assignment. In this way, the procedure branches and bounds through the paths of the OBDD until a consistent satisfying set is found or it is proved that none exists.

```
consistent_satisfy(bdd_node, var)
{
    if (var = constant ONE)
    {
        /* found satisfying set */
        justify(satisfying set);
        if (justifiable from primary inputs)
            /* found a distinguishing vector */
            return SATISFIED;
        else
            /* continue with the traversal */
            return NOT_SATISFIED;
    }
    else if (var = constant ZERO)
        /* continue with the traversal */
        return NOT_SATISFIED;

    assign:   '1' to the node in the circuit
              which the variable var represents;
    if (imply() = consistent)
        if (consistent_satisfy (var.high) = SATISFIED)
            return SATISFIED;
    erase this assignment and its implications in the circuit;

    assign:   '0' to the node in the circuit
              which the variable var represents;
    if (imply() = consistent)
        if (consistent_satisfy (var.low) = SATISFIED)
            return SATISFIED;
    erase this assignment and its implications in the circuit;
    return NOT_SATISFIED;
}
```

**Table 6.1:** Routine *consistent_satisfy()*

For the justification process *justify()* we use a test generation technique based on FAN's [FuSh83] multiple backtrace procedure and implicit enumeration. The routine *imply()* performs direct implications and makes use of the pre-stored indirect implications. Our experimental results show that *consistent_satisfy()* is, in general, reasonably efficient in solving the false negative problem. However, in many cases the process can be further speeded up considerably by the technique of the following section, which allows us to decrease the size of the OBDD traversed by *consistent_satisfy()*.

### 6.4.2 Incorporating Don't Cares

Implications can be used to incorporate partial information about don't cares in the OBDDs. The procedure is listed in Table 6.2. As pointed out, the cause of false negatives is the interdependency of the pseudo-inputs. If the equivalent nodes are independent then no false negatives can occur. Consider the function $f$ of the XOR-ed partitioned circuit with $n$ variables $x_1, x_2, ..., x_n$ that belong to the cut in the circuit, i.e., they represent pseudo-inputs of the partition for which OBDDs have been built. Function $f$ is represented as an OBDD. First $f$ is divided into two cofactors $f_i$ and $f_{\bar{i}}$ based on a variable $i \in \{ x_1, x_2, ..., x_n\}$. A value is assigned to the signal in one of the original (unpartitioned) circuits representing the variable $i$ and implications are performed. It does not matter which of the two unpartitioned circuits is taken. If the implication results in value assignments for signals representing other pseudo-inputs, then *restrictions* are made on the respective cofactors as given by Table 6.2. This process is continued for all signals. Finally, the restricted cofactors are combined by an *ITE* (If-Then-Else) operator as defined in [Brac90].

This procedure is however not complete, i.e., it does not find all don't care sets. However, experimental results show that it still reduces the OBDD sizes considerably.

**Example 6.4:** Consider the two circuits shown in Figure 6.12. The output functions $y_1$ and $y_2$ are equivalent. The OBDD for this function is shown in Figure 6.13. Note that in all examples of OBDDs the right edge of an OBDD node represents a one and the left edge represents a zero. Recursive learning with one level of recursion identifies the internal equivalent signals, $eq1$, $eq2$ and $eq3$, as shown in Figure 6.12. Based on these internal equivalences the circuits are partitioned. The partitioned circuits are shown in Figure 6.14 with their respective OBDDs. These OBDDs are not isomorphic. An XOR of these OBDDs results in an OBDD shown in Figure 6.15.

Now the procedure *consistent_satisfy* is called with this XOR-ed OBDD as argument. This procedure traverses the OBDD until a terminal one is reached. In this particular example there is only one path to a terminal one. The procedure finds this path and assigns values to corresponding nodes in the original circuits. The implications of these values lead to an inconsistency in the original circuit. The traversal is continued, but since there are no more paths that lead to a one, the two circuits are proved to be equivalent. Note that the OBDDs for the cut circuits in Figure 6.14 are considerably smaller than the OBDDs of the full circuits.

```
/*  f(x1, x2, ...,xn) is the function of the XOR-ed circuits in terms of n
    variables which belong to the pseudo-inputs of the examined
    circuit partition */

include_don't_cares(f(x1, x2, ...,xn) represented as OBDD)
{
    for (i = x1 to xn)
    {
        /* divide f into cofactors */
        Let v ∈ {0, 1}:
        f_i := |f|_{i = v};
        f_ī := |f|_{i = v̄};
        assign i := v in the original circuit;
        imply() in the original circuit;   /* make implications */
        for (j = x1 to xn and j ≠ i)
        {
            if (signal j in circuit assumes value w, w ∈ {0, 1})
                f_i := | f_i|_{j = w}
        }
        assign: i := v̄ in the original circuit;
        imply() in the original circuit;
        for (j = x1 to xn and j ≠ i)
        {
            if (signal j in circuit assumes value w, w ∈ {0, 1})
                f_ī := | f_ī|_{j = w}
        }
        f_new = ITE (i, f_i, f_ī);
        if (size (f_new < f))
            return (f_new);
        else
            return (f);
    }
}
```

**Table 6.2:** Routine *include_don't_cares()*

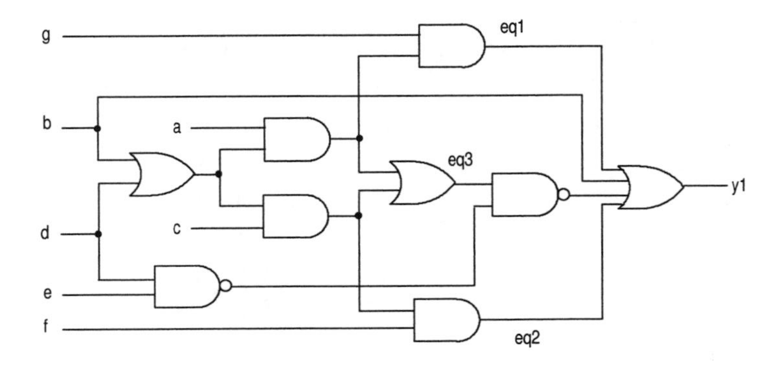


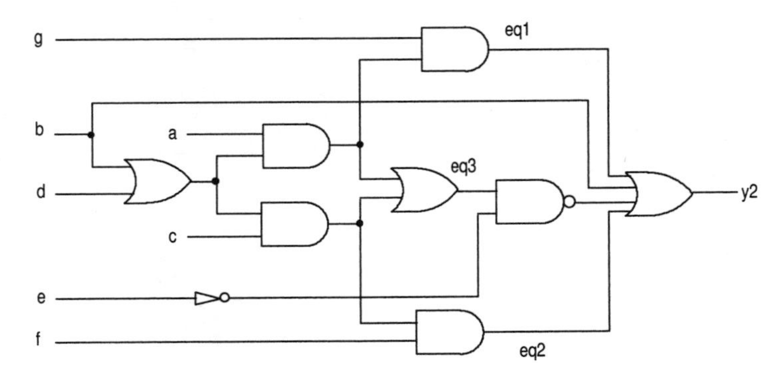


**Figure 6.12:** Example circuits (extracted from c432 [Brgl85])

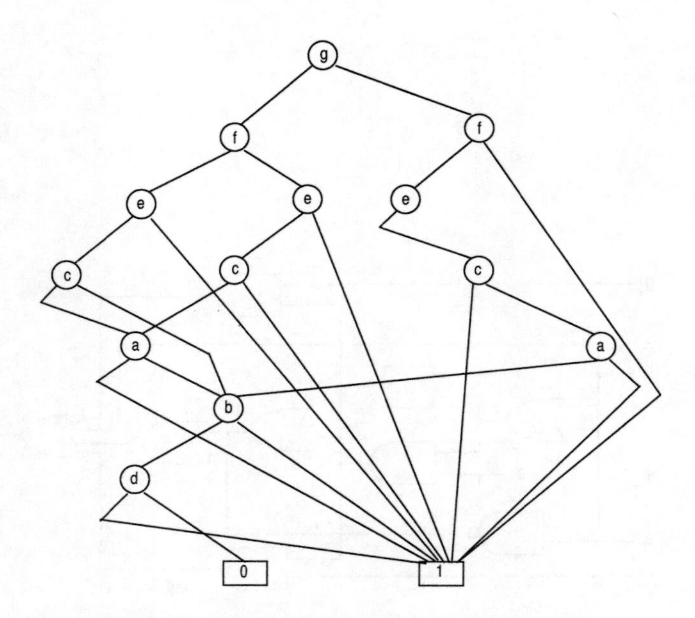


**Figure 6.13:** OBDD for circuit in Figure 6.12

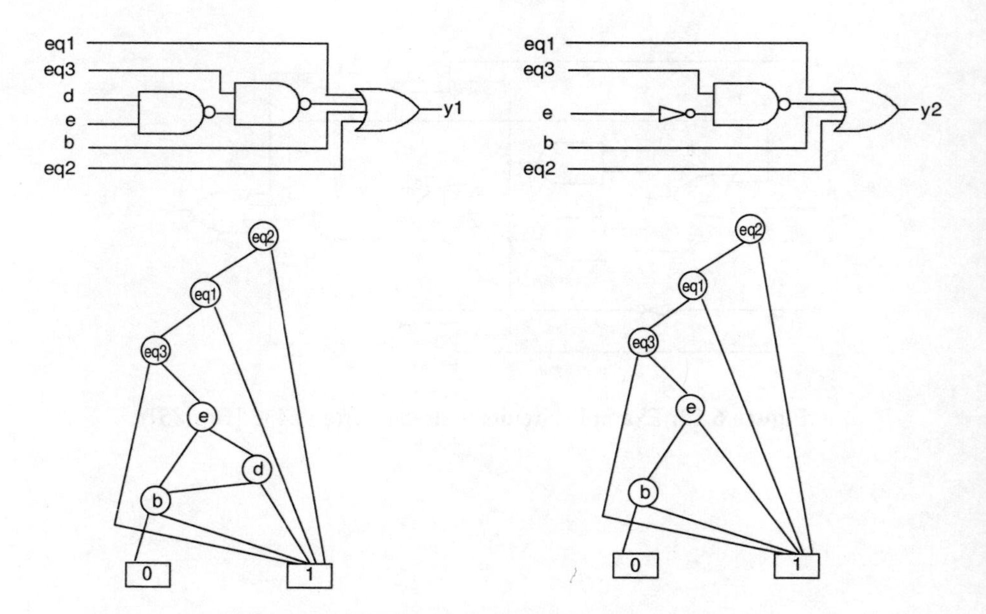


**Figure 6.14:** Partitioned circuits and their OBDDs

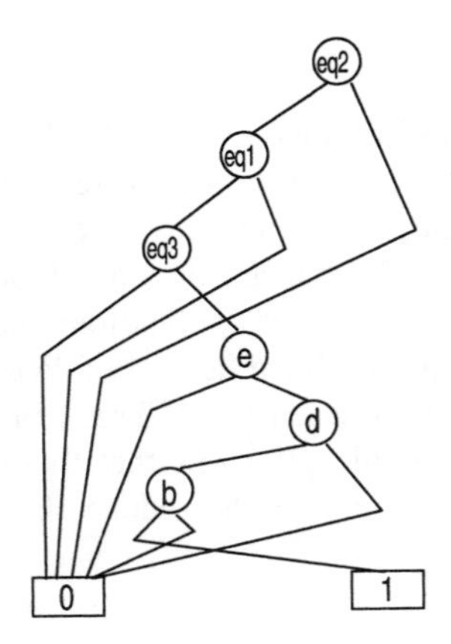


**Figure 6.15:** OBDD for XOR-ed circuit partitions (false negative)

### 6.4.3 Discussion of Possible Extensions to Functional Phase

The functional portion of the verification framework can be further improved in various ways. As new graph representations of Boolean functions become available it may be promising to examine their performance in this hybrid methodology. The experimental results described in the following section have been obtained without any sophisticated methods for variable ordering. Obviously, modern approaches to variable ordering [Rude93] can further improve these results.

In particular, experimental results suggest that more effort should be dedicated to refined solutions to the false negative problem. The method presented in Section 6.4.2 is a relatively simple example of using don't care conditions at the cut line of the circuit partitioning for OBDD minimization. More sophisticated techniques for OBDD minimization with respect to don't care conditions have been reported [ChCh94], [Ship94]. These and concepts like Coudert's *constrain* operation [CoBe89] for OBDDs seem to be useful ingredients for the described methodology. Further, note that the *compose* operation could be used to construct the circuit from the cut line backwards to the primary inputs of the original circuits. This has been investigated in [Mats96] and has led to encouraging results. If *compose* is used the techniques described above only have to be applied if the intermediate OBDD grows too large. Notice that for equivalent circuits the final OBDD is of size zero.

## 6.5 Experimental Results

In order to examine the performance of the hybrid approach a series of verification experiments has been conducted on the ISCAS-85 benchmarks and other industrial circuits. The ISCAS-85 [Brgl85] benchmarks were verified against their prime and irredundant versions [TrGo91] that are also available from MCNC. This verification experiment reflects the range of applications we have in mind for our hybrid verification method: the circuits have been modified at several different locations, but there is still "similarity" between them. This can be expected in many practical verification problems. The goal is to show that this similarity can be identified by the structural phase 1 and be used to reduce the complexity of the functional phase 2.

| | Functional | Structural + Functional | |
|---|---|---|---|
| Circuit | OBDD size | Rec. Depth | OBDD Size |
| C432 | 55,023 | 1 | 0* |
| C499 | 136,255 | 1 | 0* |
| C1355 | 136,255 | 1 | 0* |
| C1908 | 31,978 | 1 | 0* |
| C2670 | unable | 1 | 166,665 |
| C3540 | unable | 2 | 3516 |
| C5315 | 11659 | 1 | 1677 |
| C6288 | unable | 1 | 0* |
| C7552 | unable | 1 | 160,510 |

**Table 6.3:** Experimental comparison - OBDD sizes

In these experiments similarity between designs is captured by storing implications as described in Section 6.3.1. The prestored indirect implications (the internal equivalent nodes) are read from a file generated by HANNIBAL [Kunz93] in phase 1. No special variable ordering techniques are used for our BDD formation. BDD variables are created for each equivalent node based on their output distance. The ordering is fixed for all outputs of the circuits. The results are presented in Table 6.3 and Table 6.4. Table 6.3 compares the *final* OBDD sizes for the whole and the partitioned circuits, respectively. The variables in OBDDs for the whole circuit were ordered based on their output distance. Table 6.4 lists the CPU time in seconds. The recursion depth used in phase 1 of the verification algorithm is listed for each circuit in Table 6.3.

The sizes are the aggregate sizes for all outputs taking sharing into account. Importantly, in all examined cases the BDD sizes shrink drastically after the structural pre-processing phase. For some circuits marked by an * structural analysis alone completed the job [Kunz93]. For circuit c3540 we could not build a BDD for a pre-processing recursion depth of 1, so the pre-processing phase is done with a recursion depth of 2. In this way, more internal equivalences are generated which in turn makes the partitioned circuit smaller, causing the BDD sizes to shrink. This aptly demonstrates the fact that structural and functional techniques can complement each other to provide more efficient means to solve the verification problem. Note that our results can further be improved by applying more sophisticated ordering techniques reported in the literature. The BDD sizes for the examined circuits were very small compared to all conventional functional techniques. Consider the multiplier c6288. It has been proved [Brya86] that any OBDD for a multiplier grows exponentially with the number of circuit inputs so that OBDD-based verification for c6288 is not practical. In this case, the pre-processing itself has proved that the circuits are equivalent, without the need for building an OBDD.

| | Functional | Structural + Functional | | |
|---|---|---|---|---|
| Circuit | OBDD [s] | Structural [s] | Functional [s] | Total [s] |
| C432 | 60.9 | 1.0 | 0 | 1 |
| C499 | 89.32 | 1.9 | 0 | 1.9 |
| C1355 | 143.6 | 6.6 | 0 | 6.6 |
| C1908 | 30.44 | 11.2 | 0 | 11.2 |
| C2670 | unable | 8.7 | 150.6 | 159.3 |
| C3540 | unable | 52.8 | 14.84 | 67.64 |
| C5315 | 20.52 | 32.4 | 340.4 | 372.8 |
| C6288 | unable | 21.5 | 0 | 21.5 |
| C7552 | unable | 97.2 | 5486.1 | 5583.3 |

**Table 6.4:** Experimental Comparison - CPU times (Sun SPARC 5)

Table 6.5 lists the number of false negatives encountered for the benchmark circuit. The second column gives the number of outputs for each circuit. The number of outputs proved to be equivalent by the structural analysis alone are shown in the third column. The fourth and the fifth columns represent the number of outputs with isomorphic OBDDs and outputs with different OBDDs (i.e., false negatives), respectively. As the results suggest the percentage of the total number of outputs that produce false negatives is fairly low. However, efficient methods have to be incorporated to effectively deal with this problem. The method presented in Section 6.4 performed very well for all but one case of false negatives. The large CPU-time for circuit c7552 is due to one out of the ten false negatives encountered for this circuit. Some fine-tuning or an improved variable ordering for this circuit may have fixed this problem. However, we present this result because it reflects the limitation of the hybrid approach induced by the occurrence of false negatives. Fortunately, for all other cases of false negatives, also in c7552, the technique of Section 6.4 proved very efficient and contributed only slightly to the total CPU-time.

| Circuit | Total number of outputs | # outputs found equivalent by structural analysis | # outputs with isomorphic OBDDs | # outputs with different OBDDs (False Negatives) |
|---|---|---|---|---|
| C432 | 7 | 7 | - | - |
| C499 | 32 | 32 | - | - |
| C1355 | 32 | 32 | - | - |
| C1908 | 25 | 25 | - | - |
| C2670 | 140 | 126 | 6 | 8 |
| C3540 | 22 | 6 | 6 | 10 |
| C5315 | 123 | 58 | 49 | 16 |
| C6288 | 33 | 33 | - | - |
| C7552 | 108 | 56 | 42 | 10 |

**Table 6.5:** Experimental results - false negatives

The results presented so far were only for equivalent circuits. It is interesting to see how our methods fare when the circuits are not equivalent. For this reason we changed a gate in one version of the benchmark circuits and verified it against the other version. Table 6.6 presents the results for these *true negatives* and compares OBDD sizes and CPU time between the OBDD based pure functional method and our hybrid approach. For all cases the depth of recursion was one. In the majority of the cases the inequivalence was proved in the structural stage itself. This is because structural techniques are particularly powerful in generating a distinguishing vector without completely enumerating the search space. In the cases where OBDDs had to be created the required sizes were very small and in all cases the CPU time is relatively low.

These results clearly demonstrate that functional and structural methods can be combined efficiently. The complexity of functional approaches is reduced drastically if internal equivalences can be identified. Only in the worst case, i.e., if no internal equivalences exist at all, the BDDs have to be constructed to their full sizes.

| | Functional | | Structural + Functional | |
|---|---|---|---|---|
| Circuit | OBDD size | CPU time [s] | OBDD size | CPU time [s] |
| C432 | 55047 | 54.01 | 0 | 1 |
| C499 | 365,153 | 193.9 | 512 | 59 |
| C1355 | 177,573 | 285.4 | 0 | 9 |
| C1908 | 38538 | 23.27 | 0 | 10 |
| C2670 | unable | - | 0 | 11 |
| C3540 | unable | - | 10371 | 15 |
| C5315 | 12889 | 25.29 | 0 | 26 |
| C6288 | unable | - | 0 | 23 |
| C7552 | unable | - | 0 | 97 |

**Table 6.6:** Experimental comparison - true negatives

Experiments have also been conducted to evaluate the structural phase based on miter optimization as described in Section 6.3.2. Optimization based verification is useful in verifying designs that have less structural similarity than the examples considered above. We use the public domain multiplier c6288 which we verified against its optimized version. The optimized version has been obtained by SIS1.2 using *script.rugged*. The other circuits listed in Table 6.7 were obtained from Mentor Graphics Autologic II Logic Synthesis Team. These designs are highly datapath oriented and contain multipliers and rotators. They were created in Verilog and synthesized by Autologic II using a commercial ASIC vendor library. The designs were synthesized with different design goals in mind (such as area or performance). The test cases also contain intentionally non-equivalent designs. In this experiment only the structural phase of Section 6.3.2 was used. Since the circuit partitioning technique so far has only been implemented for the method of Section 6.3.1 and not for the method of Section 6.3.2, no functional methods are employed for these circuits.

The results show that the structural phase based on the optimization procedure performs efficiently and robustly for these practical verification problems. The technique of Section 6.3.2 may be outperformed by the one of Section 6.3.2 if the circuits have a high degree of similarity. On the other hand for circuits with less structural similarity as in many of these examples, logic verification by optimization provides a general framework for a more robust verification approach. Our results show that the optimization procedure of Section 5.6.2 can be tailored for efficient miter optimization. In all examined cases the miter could be minimized to a constant signal 0 within reasonable CPU-times.

| Names | # I/O | # Conn. | equivalent ? | CPU-time h : min : sec |
|---|---|---|---|---|
| c6288 with<br>c6288.rug | 32/32<br>32/32 | 4768<br>4695 | yes | 00 : 01 : 32 |
| M1.1.32 with<br>M1.2.32 | 32/16<br>32/16 | 1508<br>1508 | yes | 00 : 00 : 11 |
| M1.1.32 with<br>M1.3.32 | 32/16<br>32/16 | 1508<br>1616 | yes | 00 : 01 : 27 |
| M1.1.64 with<br>M1.2.64 | 64/32<br>64/32 | 5988<br>5988 | yes | 00 : 01 : 39 |
| M1.1.64 with<br>M1.3.64 | 64/32<br>64/32 | 5988<br>6065 | yes | 00 : 03 : 26 |
| M1.1.64 with<br>M1.4.64 | 64/32<br>64/32 | 5988<br>5981 | no | 00 : 01 : 12 |
| M2.1 with<br>M2.2 | 74/32<br>74/32 | 4318<br>4318 | no | 00 : 03 : 27 |
| M2.1 with<br>M2.3 | 74/32<br>74/32 | 4318<br>4509 | yes | 00 : 03 : 50 |
| M2.1 with<br>M2.4 | 74/32<br>74/32 | 4318<br>4508 | no | 00 : 02 : 20 |

**Table 6.7**: Logic verification with HANNIBAL (Sun SPARC 5)

As described in Section 6.3.2, miter optimization is run in two phases. The first phase performs substitution only, the second phase considers more general transformations by running redundancy elimination for the whole circuit. In all cases the first phase helped to significantly reduce the size of the miter before starting the more CPU-time expensive second phase. All circuit transformations have been derived by recursive learning with a recursion depth not greater than 4.

# Chapter 7

# CONCLUSIONS AND FUTURE WORK

This book has described Boolean reasoning techniques for *multi-level* combinational networks. In the past, a lot of research on Boolean reasoning has been conducted in the domain of two-level circuit theory. We extended the basic concept of "prime implicant" to multi-level Boolean networks and introduced AND/OR reasoning graphs for calculating these generalized prime implicants. As a special case, an implicant can consist of a single literal. Such single-literal implicants are commonly referred to as "implications" between signal values at network nodes and can be determined by a specialized AND/OR reasoning technique called recursive learning.

Complexity is the main issue when dealing with today's highly integrated circuits. We have seen that the described reasoning techniques are very useful in strategies that can avoid difficulties caused by the algorithmic complexity of problems in VLSI design automation. We have seen that single-literal implicants (implications) determine the non-solution area when solving satisfiability problems like test generation. We have described a test generation procedure that relies entirely on implications rather than on backtracking.

Formal verification is considered to be a problem of particularly high complexity. Many currently popular methods are based on some form of binary decision diagrams like OBDDs and solve the problem by representing the entire circuit function as a binary decision diagram. If the binary decision diagram can be built the problem is solvable. For some circuits, however, a BDD cannot be built. Therefore, we have described an alternative approach. Generally speaking, a complex problem can often be solved by performing a sequence of small, less complex, local operations. However, these local operations have to be selected in a smart way if they are to yield a solution of the global problem. We believe that logic implications can help derive such "smart" local operations. Applying reasoning techniques to logic equivalence checking we find that implicant-based methods can solve verification problems that cannot be solved by OBDD-based approaches alone.

Nevertheless, formal verification remains a challenging problem, especially if sequential circuits are also considered. Work is under way to extend the approach of Chapter 6 to equivalence checking for sequential circuits. Future work should also examine how *model checking* [McMi93] can take advantage of the reasoning techniques of Chapters 3 and 4. Success may be achieved by breaking the global problem into many smaller problems guided by some Boolean reasoning technique.

The methods presented in this book, to a large extent, rely on the fact that recursive learning and AND/OR reasoning graphs can represent both structural and functional properties of a circuit. Structural properties are important especially in logic synthesis. It was shown in Chapter 5 that topological properties of AND/OR reasoning graphs and certain types of implications contain important information for circuit optimization. Also here, we have seen that the reasoning techniques of Chapters 3 and 4 allow us to identify local but "smart" transformations for circuit optimization that are often superior to the transformations derived by more global BDD-based methods.

Concepts used in automatic test pattern generation (ATPG) are important ingredients for our methods. We have shown that ATPG procedures can provide the underlying "mechanics" to actually perform the transformations pointed out by the implication techniques. There is still room for improving these procedures by putting some additional filters on many operations to achieve further speed-up without sacrificing too much optimization quality [ChGi96]. Furthermore, it will be an important task for future work to adapt the described procedures for handling circuits with complex gates. Only minor extensions are needed to make the presented methods applicable to technology mapped circuit designs. Optimization on mapped netlists is appealing in practice because it allows to incorporate the specific design goals directly into the optimization procedure.

We have seen that the reasoning techniques of Chapters 3 and 4 can extract a lot of useful information about a circuit although they are restricted to relatively local considerations. Nevertheless, sometimes it would be desirable to process larger amounts of information in our reasoning techniques than we are currently able to. In fact, it should be possible by applying concepts to AND/OR reasoning graphs that are similar to those used for OBDDs. With hashing and caching techniques OBDDs can process a lot more information than ordinary Shannon trees. Similarly, perhaps AND/OR reasoning trees can be dramatically reduced in size if subtrees that contain the same prime implicants are represented only once in the graph. It should be noted however that this does not only require checking for isomorphic subtrees. In the case of AND/OR reasoning trees additional hashing rules are needed to identify all subtrees that contain equivalent information. Developping appropriate hashing techniques for AND/OR reasoning methods has been beyond the scope of this book and remains a challenge for future investigation.

# REFERENCES


[AaSt94] Aas E. J., Steen T., and Klingsheim K: "Quantifying Design Quality Through Design Experiments", *IEEE Design & Test of Computers*, vol. 11, pp. 27-38, Spring 1994.

[AbBr90] Abramovici M., Breuer M., and Friedman A.: *Digital Systems Testing and Testable Design*, Revised Printing, IEEE Press, Piscataway, New Jersey, 1994.

[AbIy92] Abramovici M., and Iyer M.: "One-Pass Redundancy Identification and Removal", *Proc. Intl. Test Conference* (ITC), pp. 807-815, 1992.

[AgBu96] Agrawal V. D., Bushnell M. L., and Lin Q.: "Redundancy Identification Using Transitive Closure", *Proc. Asian Test Symposium*, pp. 4-9, 1996.

[Aker76] Akers S.B.: "A Logic System for Fault Test Generation", *IEEE Transactions on Computers*, vol. C-25, no. 6, pp. 620-630, June 1976.

[Aker78] Akers S.: "Binary Decision Diagrams", *IEEE Transactions on Computers*, vol. C-27, no. 6, pp. 509-516, June 1978.

[Ashe59] Ashenhurst R.L.: "The Decomposition of Switching Functions", *Proc. on an International Symposium on the Theory of Switching held at Comp. Lab. of Harward University*, pp. 74-116, 1959.

[BaBr88] Bartlett K., Brayton R., Hachtel G., Jakoby R., Morrison C., Rudell R., Sangiovanni-Vincentelli A.L., and Wang A.: "Multilevel Logic Minimization Using Implicit Don't Cares", *IEEE Transactions on Computer-Aided Design*, vol. 7, no. 6, pp. 723-740, June 1988.

[BeTr89] Berman C. L., and Trevillyan L. H.: "Functional Comparison of Logic Designs for VLSI Circuits", *Proc. Intl. Conf. on Comp.-Aided Design* (ICCAD), pp. 456-459, 1989.

[BeTr91] Berman L., and Trevillyan L.: "Global Flow Optimization in Automatic Logic Design", *IEEE Transactions on Computer-Aided Design*, vol. 10, no. 5, pp. 557-564, May 1991.

[BiJa94] Bitner J, Jain J., Abadir M., Abraham J., and Fussell D.: "Efficient Algorithmic Circuit Verification Using Indexed BDDs", *Proc. Symposium on Fault-Tolerant Computing* (FTCS), pp. 266-275, 1994.

[Brac90] Brace K.: "Efficient Implementation of a BDD Package",, *Proc. Design Automations Conference* (DAC), pp. 40-45, 1990.

[Bran83] Brand D.: "Redundancy and Don't Cares in Logic Synthesis", *IEEE Transactions on Computers*, vol. C-32, no. 10, pp. 947-952, Oct. 1983.

[Bran93] Brand D.: "Verification of Large Synthesized Designs", *Proc. IEEE International Conference on Computer-Aided Design* (ICCAD), Santa Clara, pp. 534-537, Nov. 1993.

[Bran94] Brand D. et al.: "Incremental Synthesis", *Proc. IEEE International Conference on Computer-Aided Design* (ICCAD), San Jose, pp. 14-18, Nov. 1994.

[Bray84] Brayton R. K., Hachtel G. D., McMullen C. T., and Sangiovanni-Vincentelli A. L.: *Logic Minimization Algorithms for VLSI Synthesis*, Kluwer Academic Publishers, Boston, 1984.

[Bray87] Brayton R. K., Rudell R., Sangiovanni-Vincentelli A., and Wang A. R.: "MIS: Multi-level Interactive Logic Optimization System", *IEEE Transactions on Computer-Aided Design*, vol. 6, no. 6, pp. 1062-1081, Nov. 1987.

[Brgl83] Brglez F.: "Testability in VLSI", *Proceedings 1983 Canadian Conference on VLSI*, October 1983.

[Brgl85] Brglez F., and Fujiwara H.: "A Neutral Netlist of 10 Combinational Benchmark Designs and a Special Translator in Fortran", *Proc. Intl. Symp. on Circuits and Systems* (ISCAS), Special Session on ATPG and Fault Simulation, June 1985.

[Brgl89] Brglez F. et al.: "Combinational Profiles of Sequential Benchmark Circuits", *Proc. Intl. Symp. on Circuits and Systems* (ISCAS), pp. 1929-1934, May 1989.

[BrMu82] Brayton R., and McMullen C.: "The Decomposition and Factorization of Boolean Expressions", *Proc. Intl. Symposium on Circuits and Systems* (ISCAS), pp. 49-54, 1982.

[Brow90] Brown F.: *Boolean Reasoning*, Kluwer Academic Publishers, Boston, MA, 1990.

[BrSe88] Brayton R.K., Sentovich E.M., and Somenzi F.: "Don't Cares and Global Flow Analysis of Boolean Networks", *Proc. IEEE International Conference on Computer-Aided Design* (ICCAD), pp. 98-101, 1988.

[Brya86] Bryant R.: "Graph-based algorithms for Boolean function manipulation", *IEEE Transactions on Computers*, vol. C-35, no. 8, pp. 677-691, August 1986.

[CeMa90] Cerny E., and Mauras C.: "Tautology Checking Using Cross-Controllability and Cross-Observability Relations", *Proc. Intl. Conference on Computer-Aided Design* (ICCAD), pp. 34-38, 1990.

[ChAg91] Chakradhar S.T., and Agrawal V.D.: "A Transitive Closure based Algorithm for Test Generation", *Proc. 28th Design Automation Conference* (DAC), pp. 353-358, 1991.

[ChAg93] Chakradhar S.T., Agrawal V.D., and Rothweiler S.: "A Transitive Closure Algorithm for Test Generation", *IEEE Transactions on Computer-Aided Design*, vol. 12, no. 7, pp. 1015-1028, July 1993.

[ChBu91] Chakradhar S.T., Agrawal V.D., and Bushnell M.L.: *Neural Models and Algorithms for Digital Testing*, Kluwer Academic Publishers, Boston, 1991.

[ChBu96] Chen X., and Bushnell M.: *Efficient Branch and Bound Search with Application to Computer-Aided Design*, Kluwer Academic Publishers, Boston, 1996.

[ChCh94] Chang S.-C., Cheng D.I., and Marek-Sadowska M.: "Minimizing ROBDD Size of Incompletely Specified Multiple Output Functions", *Proc. European Design Automation Conference* (EDAC), pp. 620-624, 1994.

[Chen94] Cheng K.T., April 1994, private communication.

[ChGi96] Chang S.H., van Ginneken L., and Marek-Sadowska M.M.: "Fast Boolean Optimization by Rewiring", *Proc. International Conference on Computer-Aided Design* (ICCAD), pp. 262-269, Nov. 1996.

[ChMa94] Chang S.C., and Marek-Sadowska M.: "Perturb and Simplify: Multi-Level Boolean Network Optimizer", *Proc. International Conf. on Computer-Aided Design* (ICCAD), San Jose, pp. 2-5, Nov. 1994.

[ChPr95] Chatterjee M., Pradhan D., and Kunz W.: "LOT: Optimization with Testability - New Transformations using Recursive Learning", *Proc. International Conf. on Computer-Aided Design* (ICCAD), San Jose, pp. 318-325, Nov. 1995.

[CoBe89] Coudert O., Berthet C., and Madre J.C.: "Verification of Synchronous Sequential Machines Based on Symbolic Execution", in *Lecture Notes in Computer Science*, vol. 407, (Automatic Verification Methods for Finite State Systems, International Workshop, Grenoble, France), Springer-Verlag, June 1989.

[Curt61] Curtis H.A.: "A Generalized Tree Circuit", *J. Assoc. Comput. Mach.*, pp. 484-496, Aug. 1961.

[DaYa95] Damiani M., Yang C.Y., and De Micheli G.: "Optimization of Combinational Logic Circuits Based on Compatible Gates", *IEEE Transactions on Computer-Aided Design*, vol. 14, no. 11, pp. 1316-1328, Nov. 1995.

[DeMi94] De Micheli G.: *Synthesis and Optimization of Digital Circuits*, McGraw-Hill, 1994.

[Drec94] Drechsler R., Sarabi A., Theobald M., Becker B. and Perkowski M.: "Efficient Representation and Manipulation of Switching Functions Based on Ordered Kronecker Functional Decision Diagrams", *Proc. Design Automation Conference* (DAC) pp. 415-419, 1994.

[EnCh93] Entrena L. A., and Cheng K.T: "Sequential Logic Optimization by Redundancy Addition and Removal", *Proc. Intl. Conf. on Computer-Aided Design* (ICCAD), pp. 310-315, Nov. 1993.

[Fabr92] Fabricius E.: *Modern Digital Design and Switching Theory, CRC Press*, 1992.

[FuFu92] Fujino T., and Fujiwara H.: "An Efficient Test Generation Algorithm Based on Search State Dominance", *Proc. Intl. Symp. on Fault-Tolerant Comp.* (FTCS), pp. 246-253, 1992.

[FuKa91] Fujita M., Kakuda T., and Matsunaga Y.: "Redesign and Automatic Error Correction of Combinational Circuits", Logic and Architecture Synthesis, ed. G. Saucier, North-Holland: Elsvier Science Publishers B.V., pp. 253-262, 1991.

[FuMa91] Fujita M., Matsunaga Y., and Kakuda T.: "On Variable Ordering of Binary Decision Diagrams for the Application of Multi-Level Logic Synthesis", *Proc. European Design Automation Conference*, pp. 50-54, March 1991.

[FuSh83] Fujiwara H., and Shimono T.: "On the Acceleration of Test Generation Algorithms", *Proc. 13th Int. Symp. on Fault Tolerant Computing* (FTCS), pp. 98-105, 1983.

[GeBr88] McGeer P., and Brayton R.K.: "Efficient, Stable Algebraic Operation on Boolean Expressions", in C. Sequin (Ed.) *VLSI Design of Digital Systems*, North-Holland, Amsterdam, 1988.

[GeBr89] McGeer P., and Brayton R.K.: "Consistency and Observability Invariance in Multi-Level Logic Synthesis", *Proc. Int. Conf. on Computer-Aided Design* (ICCAD), pp. 426-429, 1989.

[GiBu91] Giraldi, J., and Bushnell M.: "Search State Equivalence for Redundancy Identification and Test Generation", *Proc. Intl. Test Conf.* (ITC), pp. 184-193, 1991.

[Goel81] Goel P.: "An Implicit Enumeration Algorithm to Generate Tests for Combinational Logic Circuits", *IEEE Transactions on Computers*, vol. C-30, no. 3, pp. 215-222, March 1981.

[Gold80] Goldstein L., and Thigpen E.: "SCOAP: Sandia Controllability / Observability Analysis Program" *Proceedings 17th Design Automation Conference* (DAC), pp. 190-196, June 1980.

[HaJa88] Hachtel G., Jacoby R., Moceyunas P., and Morrison C.: "Performance Enhancements in BOLD using Implications", *Proc. Intl. Conf. on Computer-Aided Design* (ICCAD), pp. 94-97, Nov. 1988.

[HaSo96] Hachtel G., and Somenzi F.: *Logic Synthesis and Verification Algorithms*, Kluwer Academic Publishers, Boston 1996.

[HeWi95] Henftling M., Wittmann H., and Antreich K.J.: "A Formal Non-Heursitic ATPG Approach", *Proc. European Design Automation Conference* (EURODAC), pp. 248 -253, 1995.

[HwOw90] Hwang T., Owens R.M., and Irwin M.J.: "Exploiting Communication Complexity for Multi-Level Logic Synthesis", *IEEE Transactions on Computer-Aided Design*, vol. 9, no. 10, pp. 1017-1027, Oct. 1990.

[IbSa75] Ibarra O.H., and Sahni S.K.: "Polynomially complete fault detection problems", *IEEE Transactions on Computers*, vol. C-24, pp. 242-249, March 1975.

[IsSa91] Ishiura N., Sawada H., and Yajima S.: "Minimization of Binary Decision Diagrams Based on Exchanges of Variables", *Proc. Intl. Conf. on Computer-Aided Design* (ICCAD), pp. 472-475, 1991.

[JaAg85] Jain S.K., and Agrawal V.D., "Statistical Fault Analysis", *IEEE Design and Test of Computers*, vol. 2, pp. 38 - 44, Feb. 1985.

[JaMu95] Jain J., Mukherjee R., and Fujita M.: "Advanced Verification Techniques Based on Learning", *Design Automation Conference* (DAC), pp. 420 - 426, June 1995.

[Karp63] Karp R.: "Functional Decomposition and Switching Circuit Design", *J. Soc. Indust. Appl. Math.*, vol. 11, no. 2, pp. 291-335, 1963.

[KeSc92] Kebschull U., Schubert E., and Rosenstiel W.: "Multi-level Logic Based on Functional Decision Diagrams", *Proc. European Design Automation Conf.* (EDAC), pp. 43-47, 1992.

[KiMe87] Kirkland T., and Mercer M.R.: "A Topological Search Algorithm For ATPG", *Proc. 24th Design Automation Conference* (DAC), pp. 502-508, June 1987.

[Koha78] Kohavi Z.: *Switching and Finite Automata Theory*, McGraw-Hill, New York, 1978.

[KrWu90] Kropf T., and Wunderlich H.: "A Common Approach to Test Generation and Hardware Verification Based on Temporal Logic", *Proc. International Test Conference* (ITC), pp. 57-66, 1990.

[KuMe94] Kunz W., and Menon P.: "Multi-Level Logic Optimization by Implication Analysis", *Proc. Intl. Conference on Computer-Aided Design* (ICCAD), San Jose, pp. 6-13, Nov. 1994.

[Kund92] Kundu S. et al.: "A Small Test Generator for Large Designs", *Proc. Intl. Test Conf.* (ITC), pp. 30-40, 1992.

[Kunz93] Kunz W.: "HANNIBAL: An Efficient Tool for Logic Verification Based on Recursive Learning", *Proc. Intl. Conference on Computer-Aided Design* (ICCAD), Santa Clara, pp. 538-543, Nov. 1993.

[KuPr92] Kunz W., and Pradhan D.: "Recursive Learning: An Attractive Alternative to the Decision Tree for Test Generation in Digital Circuits", *Proc. Intl. Test Conference* (ITC), pp. 816-825, 1992.

[KuPr93] Kunz W., and D. Pradhan: "Accelerated Dynamic Learning for Test Generation in Combinational Circuits", *IEEE Transactions on Computer-Aided Design*, vol. 12, no. 5, pp. 684-694, May 1993.

[KuPr94] Kunz W., and Pradhan D.K.: "Recursive Learning: A New Implication Technique for Efficient Solutions to CAD Problems: Test, Verification and Optimization", *IEEE Transactions on Computer-Aided Design*, vol. 13, no. 9, pp. 1143-1158, Sept. 1994.

[KuPr96] Kunz W., Pradhan D., and Reddy S.: "A Novel Framework for Logic Verification in a Synthesis Environment", *IEEE Transactions on Computer-Aided Design*, vol. 15, no. 1, pp. 20-33, January 1996.

[KrWu90] Kropf T., and Wunderlich H.: "A Common Approach to Test Generation and Hardware Verification Based on Temporal Logic", *Proc. International Test Conference* (ITC), pp. 57-66, 1990.

[LaPa96] Lai Y., Pan K.R., and Pedram M.: "OBDD-Based Function Decomposition: Algorithms and Implementation", *IEEE Transactions on Computer-Aided Design*, vol. 15, no. 8, pp. 977-990, August 1996.

[LaPe93] Lai Y., Pedram M., and Vrudhula S.: "BDD Based Decomposition of Logic Functions with Application to FPGA Synthesis," *Proc. Design Automation Conference* (DAC), pp. 642-647, June 1993.

[Larr89] Larrabee T.: "Efficient Generation of Test Patterns Using Boolean Difference", *Proc. Intl. Test Conf.* (ITC), pp. 795-801, 1989.

[Lawl64] Lawler E.: "An Approach to Multilevel Boolean Minimization", *Journal of the ACM*, vol. 11, pp. 283-295, 1964.

[LeHa91] Lee H.K., and Ha D.S.: "An Efficient Forward Fault Simulation Algorithm Based on the Parallel Pattern Single Fault Propagation", *Proc. Intl. Test Conference* (ITC), pp. 946-953, Sept, 1991.

[Lioy87] Lioy A., and Mezzalama M.: "On Parameters Affecting ATPG Performance", *Proceedings of CompEuro* 1987, May 1987.

[MaGr90] Mahlstedt U., Grüning T., Özcan C., and Daehn W.: "CONTEST: A Fast ATPG Tool for Very Large Combinational Circuits", *Proc. Intl. Conference on Computer Aided Design* (ICCAD), pp. 222-225, Nov. 1990.

[Matr90] Matrosova A.: "Algorithmic Methods for Testing", *Tomsk State University publisher*, Tomsk, Russia, 1990.

[Mats96] Matsunaga Y.: "An efficient equivalence checker for combinational circuits", *Proc. Design Automation Conference*, pp. 629-634, Las Vegas, 1996.

[MaWa88] Malik S., Wang A., Brayton R., and Sangiovanni-Vincentelli A.: "Logic Verification Using Binary Decision Diagrams in a Logic Synthesis Environment", *Proc. Int. Conf. on Computer-Aided Design* (ICCAD), pp. 6-9, Nov. 1988.

[McCl56] McCluskey E.: "Minimization of Boolean Functions", *Bell System Technical Journal*, vol. 35, pp. 1417-1444, 1956.

[McCl86] McCluskey E.: *Logic Design Principles*, Prentice-Hall, Englewood Cliffs, NJ, 1986.

[McMi93] McMillan K.: *Symbolic Model Checking*, Kluwer Academic Publishers, Boston 1993.

[MoSc93] Molitor P., and Scholl C.: "Communication Based Multilevel Synthesis for Multi-Output Boolean Functions", *Proc. 4th Great Lakes Symposium on VLSI*, pp. 101-104, 1994.

[MuBr94] Murgai R., Brayton R., and Sangiovanni-Vincentelli A.: "Optimum Functional Decomposition Using Encoding", *Proc. Design Automation Conf.* (DAC), pp. 408-414, 1994.

[Muro89] Muroga S. et al.: "The Transduction Method - Design of Logic Networks Based on Permissible Functions", *IEEE Transactions on Computers*, vol. C-38, no. 10, pp. 1404-1424, Oct. 1989.

[Muth76] Muth P.: "A Nine-Valued Logic Model for Test Generation", *IEEE Transactions on Computers*, vol. C-25,no. 6, pp. 630-636, June 1976.

[PoRe93] Pomeranz I., and Reddy S.M.: "On Diagnosis and Correction of Design Errors", *Proc. of Intl. Conf. on Computer-Aided Design* (ICCAD), pp. 500-507, 1993.

[PoRe96] Pomeranz I., and Reddy S.M.: "On Removing Redundancies from Synchronous Sequential Circuits with Synchronizing Sequences", *IEEE Transactions on Computers*, vol. 45, no. 1, pp. 20-33, January 1996.

[PrCh96] Pradhan D., Chatterjee M., Swarna M., and Kunz W.: "Implication-Based Gate-Level Synthesis for Low-Power", *Proc. IEEE Symp. of Low Power Electronics and Design*, pp. 297-300, 1996.

[Quin52] Quine W.: "The Problem of Simplifying Truth Functions", *American Mathematical Monthly*, vol. 59, pp. 521-531, 1952.

[RaCo90] Rajski J., and Cox H.: "A Method to Calculate NecessaryAssignments in Algorithmic Test Pattern Generation", *Proc., Int. Test Conf.* (ITC), pp. 25-34, 1990.

[RaVa90] Rajski J., and Vasudevamurthy J.: "Testability Preserving Transformations in Multi-Level Logic Synthesis", *Proc. Intl. Test Conference* (ITC), pp. 265-273, 1990.

[ReKu95] Reddy S., Kunz W., and Pradhan D.: "A Novel Verification Framework Combining Structural and OBDD Methods in a Synthesis Environment", *Proc. Design Automation Conference* (DAC), pp. 414-419, June 95.

[Rich83] Rich E.: *Artificial Intelligence*, McGraw-Hill, 1983.

[Robi65] Robinson, J.A.: "A Machine Oriented Logic Based on the Resolution Principle", *Journal of the Association for Computing Machinery*, vol. 12, no. 1, pp. 23-41, January 1965.

[RoBr94] Rohfleisch B., and Brglez F.: "Introduction of Permissible Bridges with Application to Logic Optimization after Technology Mapping", *Proceedings EDAC/ETC /EUROASIC*, pp. 87-93, 1994.

[RoKa62] Roth J.P., and Karp R.M.: "Minimization over Boolean Graphs", *IBM Journal of Research and Development*, vol. 6, no. 2, pp. 227-238, April 1962.

[Roth66] Roth J.P.: "Diagnosis of Automata Failures: A Calculus and a Method", *IBM Journal of Research and Development*, vol. 10, no. 4, pp. 278-291, July 1966.

[RoWu95] Rohfleisch B., Wurth B., and Antreich K.: "Logic Clause Analysis for Delay Optimization", *Proc. Design Automation Conference* (DAC), pp. 668-672, 1995.

[Rude89] Rudell R.: "Logic Synthesis for VLSI Design", Memorandum UCB/ERL M89/49, Ph.D. Dissertation, University of California at Berkeley, April 1989.

[Rude93] Rudell R.: "Dynamic Variable Ordering for Ordered Binary Decision Diagrams", *Proc. Intl. Conf. on Computer-Aided Design* (ICCAD), pp. 42-47, 1993.

[SaBr91] Savoj H., Brayton, R.K., and Touati H.: "Extracting Local Don't Cares for Network Optimization", *Proc. Intl. Conf. on Computer-Aided Design* (ICCAD), pp. 514-517, Nov. 1991.

[Sasa93] Sasao T. (ed.): *Logic Synthesis and Optimization*, Kluwer Academic Publishers, Boston, MA, 1993.

[ScAu89] Schulz M., and Auth E.: "Improved Deterministic Test Pattern Generation with Applications to Redundancy Identification", *IEEE Trans. on Computer-Aided Design*, vol. 8, no. 7, pp. 811-816, July 1989.

[ScTr87] Schulz M., Trischler E., and Sarfert T.: "SOCRATES: A highly efficient automatic test pattern generation system", *Proc. Intl. Test Conf.* (ITC), pp. 1016-1026, 1987.

[ScTr88] Schulz M., Trischler E., and Sarfert T.: "SOCRATES: A highly efficient automatic test pattern generation system", *IEEE Transactions on Computer-Aided Design*, vol. 7, no.1, pp. 126-137, January 1991.

[Shan38] Shannon C.: "A Symbolic Analysis of Relay and Switching Circuits", *Transactions AIEE*, vol. 57, pp. 713-723, 1938.

[ShDe93] Shen A., Devadas S., and Ghosh A.: "Probabilistic Construction and Manipulation of Free Boolean Diagrams", *IEEE International Conference on Computer-Aided Design* (ICCAD), pp. 544-549, Santa Clara, Nov. 1993.

[Ship94] Shiple T. R.: "Heuristic Minimization of BDDs Using Don't Cares", *Proc. Design Automation Conference* (DAC), pp. 225-231, 1994.

[StBh91] Stanion T., and Bhattacharya D.: "TSUNAMI: A path-oriented scheme for algebraic test generation", *Proc. Fault-Tolerant Computing Symposium* (FTCS), pp. 36-43, June 1991.

[StKu97] Stoffel D., Kunz W., and Gerber S.: "AND/OR Reasoning Graphs for Determining Prime Implicants in Multi-Level Combinational Networks", *Proc. Asia and South Pacific Design Automation Conference* (ASP-DAC), Tokyo, Japan, pp. 529-538, Jan. 1997.

[StSt94] Steinbach B., and Stöckert M.: "Design of Fully Testable Circuits by Functional Decomposition and Implicit Test Pattern Generation", *Proc. 12th IEEE VLSI Test Symposium*, pp. 22-27, April 1994.

[TrGo91] Tromp G. J., and van de Goor A. J.: "Logic Synthesis of 100-percent Testable Logic Networks", *IEEE Intl. Conference on Computer-Design* (ICCD), pp. 428-431, Sept. 1991.

[TrJo86] Trevillyan L., Joyner and W., Berman L.: "Global Flow Analysis in Automatic Logic Design" *IEEE Transactions on Computers*, vol. C-35, no. 1, pp. 77-81, January 1986.

[Ubar76] Ubar R.: "Test Generation for Digital Circuits Using Alternative Graphs", *Proc. of Tallinn Technical University*, Estonia, No. 409, pp. 75-81, 1976, in Russian.

[Ubar94] Ubar R.: "Test Generation for Digital Systems Based on Alternative Graphs", *Proc. Dependable Computing-EDCC-1*, Springer Lecture Notes in Computer Science, pp. 151-164, 1994.

[WaGu96] Watanabe Y., Guerra L., and Brayton R.: "Permissible Functions for Multiouput Components in Combinational Logic Optimization", *IEEE Transactions on Computer-Aided Design*, vol. 15, no. 7, pp. 732-744, July 1996.

[WaSh90] Waicukauski J.A., Shupe P.A., Giramma D.J., and Matin A.: "ATPG for Ultra-Large Structured Designs", *Proc. Intl. Test Conf.* (ITC), pp. 44-51, 1990.

[Wege87] Wegener I.: "The Complexity of Boolean Functions", *B.G. Teubner Stuttgart*, 1987.

[WuEc95] Wurth B., Eckl K., and Antreich K.: "Functional Multiple-Output Decomposition: Theory and an Implicit Algorithm", *Proc. Design Automation Conf.* (DAC), pp. 54-59, June 1995.

[YuNa95] Yuguchi M., Nakamura Y., Wakabayashi K., and Fujita T.: "Multi-Level Logic Minimization based on Multi-Signal Implications", *Proc. Design Automation Conf.* (DAC), pp. 658-662, 1995.

# APPENDIX

**Theorem 3.1:** The procedure in Table 3.3 makes complete implications; i.e., a finite $r_{final}$ always exists such that *make_all_implications(0, $r_{final}$)* determines all necessary assignments for a given set of value assignments with unjustified gates, $U^0$. If the initial set of value assignments is impossible in the circuit, then a logic inconsistency is produced.

**Proof:**

*Notation:*

${}^{g}J^{r} = \{f_1=V_1, f_2=V_2, \ldots\}$ is a set of assignments in level $r$ that represents a justification for some gate g in the recursion level $r$-1.

${}^{g}C^{r} = \{J_1, J_2, J_3, \ldots\}$ is a complete set of justifications for gate g in the recursion level $r$.

${}^{J}U^{r} = \{g_1, g_2, g_3, \ldots\}$ is a set of unjustified gates in recursion level $r$ as determined from a given justification $J$.

*Preliminary Remarks:*

A crucial point to realize about the recursive learning procedure is that by examining justifications at unjustified gates *separately* we draw conclusions which are valid for the objective that the unjustified gates have to be justified *simultaneously*.

**Definition P.1:** Let $U$ be the set of unjustified gates for some initial set of value assignments. A justification $J$ for some unjustified gate $g \in U$ is called logically consistent, if there exists a set of value assignments in the circuit called *satisfying set S* for $J$ with the following properties:

i) $S$ justifies $g$

ii) $S$ is associated with a set of unjustified gates $U'$, with $U' \subseteq U$, and for each gate $g' \in U'$ there is a complete set of justifications ${}^{g'}C'$ being identical to the corresponding ${}^{g}C$ in $U$.

This definition of a satisfying set is only used in this proof and differs from what normally may be called a satisfying set. Note that not all unjustified gates have to be justified in this definition of a satisfying set.

Suppose we are given a set of value assignments $I$ and for each $g \in U$ there exists at least one consistent justification (for each consistent justification there is at least one satisfying set $S$.) Further, suppose we pick a gate $g$ and an arbitrary satisfying set $S$ for $g$. If we make assignments from $S$ then, by Definition P.1, we are left with a subset $U'$ of $U$ of unjustified gates and there are no other unjustified gates. Next we pick a gate $g' \in U'$, choose a satisfying set $S'$ for $g'$ and make assignments. Because $S'$ is a satisfying set these assignments do not produce a conflict with any unjustified gate in $U$ and since there are no other unjustified gates (by definition of the satisfying set) no conflict can occur. We are now left with a set $U''$, $U'' \subseteq U' \subseteq U$. This can be continued until the set of unjustified gates becomes empty. Hence, if a satisfying set according to Definition P.1 exists for every unjustified gate, i.e., if a consistent justification exists for every unjustified gate, then a set of value assignments exists that justifies *all* unjustified gates in the circuit *simultaneously* and hence the given situation of value assignments is consistent. If there is an unjustified gate where no satisfying set exists then the set of value assignments is not consistent.

Hence, by Definition P.1, a situation of value assignments is consistent if and only if there exists at least one consistent justification (separately) for every unjustified gate. Similarly we can conclude that an assignment is necessary to simultaneously satisfy all unjustified gates from a given set of value assignments if and only if it is contained in every satisfying set of at least one unjustified gate, i.e., if it occurs in every consistent justification of at least one unjustified gate.

In a given level of recursion, procedure *make_all_implications()* checks whether there is a consistent justification for every unjustified gate and extracts all assignments common to all consistent justifications. Therefore, with the above, in a given level of recursion, the procedure will correctly determine the consistency of a given set of value assignments and identify all necessary assignments if for each justification we know

a) whether or not the justification is consistent and
b) all assignments that are necessary for the consistency of the justification.

Now the theorem can be proved by induction.

*Induction:*

1) Take $r = final\text{-}1$ :

As more recursions are performed, more assignments are made; i.e., for all unjustified gates in $U^0$, the recursive call of *make_all_implications()* will always reach a level *final,* such that ${}^{J}U^{final} = \varnothing$ for all justifications $J_i \in {}^{g}C^{final\text{-}1}$ and for all gates $g$ in $U^{final\text{-}1}$. This is the case when implications have reached primary inputs or outputs or they get absorbed in unjustified gates which they justify.

The set of value assignments induced by a justification in the level *final* is a satisfying set by Definition P.1 because no new unjustified gates are created. Therefore, in the level *final* we can speak of consistent justifications and satisfying sets interchangeably. For the set of value assignments in the level *final*-1 we determine that:

a) it is consistent if and only if for every unjustified gate there is at least one consistent justification, and

b) a value assignment $f = V$, where $V \in B$ and $B$ is some logic alphabet, is necessary for level *final*-1 if and only if there exists a gate $g \in U^{final\text{-}1}$ such that $f = V$ is contained in every consistent justification of $g$.

This applies to any set of value assignments in the level *final*-1 resulting from some justification $J_i^{final\text{-}1} \in C^{final\text{-}2}$ for arbitrary $g \in U^{final\text{-}2}$. Hence, for the level *final*-1 we know the logic consistency and all necessary assignments for any set of value assignments resulting from some justification $J_i^{final\text{-}1} \in C^{final\text{-}2}$ for arbitrary $g \in U^{final\text{-}2}$.

2) Assume that,

we are in level $n$-1 and that we know the logic consistency and all necessary assignments for all sets of value assignments in level $n$ resulting from arbitrary justifications $J_i^n \in C^{n\text{-}1}$ for arbitrary $g \in U^{n\text{-}1}$.

3) Then,

with the preliminary remarks we conclude that procedure *make_all_implications()* will correctly determine the consistency and identify all necessary assignments for every set of value assignments in level $n$-1. By complete induction we conclude that we obtain all necessary assignments for the initial set of value assignments with unjustified gates $U^0$.

q.e.d.

**Theorem 4.1:** Let $y$ be the output signal of a two-level combinational circuit in SOP-form. The AND/OR tree for the assignment $y = 0$ (tautology test) has only 2 levels if the SOP-expression is unate.

**Proof:**

The completeness of routine *and_or_enumerate()* results immediately from Theorem 3.1. Then, the theorem follows from the following arguments: The output signals of the implicants in the SOP become unjustified lines after the value assignments $y = 0$. Their justifications reach the primary inputs. The implications from the justifications may cause events at other implicants (unjustified lines). However, since the SOP is unate the implications from the justifications will produce values that justify these unjustified lines so that no new AND nodes can be created.

q.e.d.

**Theorem 4.2:** Let $y$ be an arbitrary node in a combinational network and $T$ be the AND/OR enumeration tree for an initial set of value assignments $S = \{y = 0\}$. Consider a product term $t = x_1 \cdot x_2 \cdot ...x_m$ where $x_i$ is a literal corresponding to a variable $f_i$ or its complement in the combinational network. Further, consider an IST of $T$ with a set of leaves $L$.

If there is a one-to-one mapping between the literals $x_i$ of $t$ and the elements $(f_i = V_i)$ of $L$ such that $V_i = 0$ if $x_i$ represents the uncomplemented variable $f_i$, and $V_i = 1$ if $x_i$ represents the complemented variable $\overline{f_i}$, then $t$ is a 1-implicant for $y$. Analogously, $t$ is a 0-implicant for $y$ if the IST is a subtree of the enumeration tree with the initial assignment $S = \{y = 1\}$.

**Proof:**

The theorem is "obvious" due to the structure of the AND/OR tree. Nevertheless, for reasons of completeness it is proved formally by noting that the AND/OR tree is isomorphic to a Boolean expression of the recursive form described below:

Let $n$ be the level index associated with the nodes of the AND/OR tree and $o$ and $a$ be Boolean expressions associated with the OR nodes and AND nodes such that the $o_j$ are the children of the $a_i$ and the $a_k$ are the children of the $o_j$ :

$$^{n}a_i = \prod_j {}^{n}o_j \qquad (1)$$

$$^{n}o_j = \begin{cases} \sum_k {}^{n+1}a_k & \text{for non-terminal nodes} \qquad (2) \\ \bar{f} & \text{for terminal nodes if leaf corresponds to } f = 0 \\ f & \text{for terminal nodes if leaf corresponds to } f = 1 \end{cases}$$

By recursively applying Equations (1) and (2) to all levels $n$ we obtain a Boolean expression $^{0}a$ with:

$$y = 0 \Rightarrow {}^{0}a = 1 \qquad (3)$$

The proof is by *induction*:

Consider an implication subtree (IST) of the AND/OR enumeration tree for the initial value assignment $S = \{y = 0\}$.

1) Take: the leaves of the IST, $^{final}o_j$

   for all leaves we set $^{final}o_j = 0$. This means that the product term $t$ formed by the leaves of the IST as given in the theorem evaluates to 1. Then, since in the IST there exists at least one OR child of each AND node, for every $^{final}a_i$ there exists a $^{final}o_j$ such that $^{final}o_j = 0$.

2) Assumption:

   for every $^{n}a_i$ there exists a $^{n}o_j$ child such that $^{n}o_j = 0$.

3) Then, for all $^{n-1}a_i$ it is $^{n-1}a_i = 0$

   Proof:
   given the above assumption,
   Eq. (1) $\Rightarrow$ for every $^{n}a_i$ in the IST it is $^{n}a_i = 0$,
   for a given $^{n}a_i$ all its siblings are also included and are 0, and with
   Eq.(2) $\Rightarrow$ for every $^{n-1}o_j$ in the IST it is $^{n-1}o_j = 0$,
   Eq. (1) $\Rightarrow$ for every $^{n-1}a_i$ in the IST it is $^{n-1}a_i = 0$.

By induction we conclude that the Boolean expression $^{0}a$ belonging to the IST becomes 0 if the product term $t$ formed as given in the theorem evaluates to 1, i.e.,

$$t = 1 \Rightarrow {}^{0}a = 0 \qquad (4)$$

By contraposition of (3) we conclude with (4) that $t = 1 \Rightarrow y = 1$, and hence $t$ is a 1-implicant of $y$. The proof for a 0-implicant is analogous.

q.e.d.

**Theorem 4.3:** Let $y$ be an arbitrary node in a combinational network and $T$ be the AND/OR reasoning tree for an initial set of value assignments $S = \{y = V\}$, $V \in \{0, 1\}$. For every *prime* implicant of $y$ there exists a minimal implication subtree (MIST) of $T$ such that the leaves of the MIST correspond to the literals of the prime implicant as given in Theorem 4.2.

**Proof:**

Routine *make_all_implications()*, described in Section 3.2.1 can identify all single-literal implicants for a function $y$ in a combinational network. This is equivalent to performing all logic implications of $y = 0$. It is accomplished by recursively checking whether all consistent justifications contain the same value assignments. Viewed in the AND/OR tree this corresponds to checking the properties of an IST as defined in Definition 4.5 such that the IST has identical leaves. Clearly, in the AND/OR enumeration tree single literal implicants belong to implication subtrees with identical leaves. Further, since *make_all_implications()* has been proved complete, every single-literal implicant can be associated with an IST having identical leaves.

The theorem is now proved for a prime 1-implicant using the following construction: Suppose there is a product term $t = x_1 \cdot x_2 \cdot \ldots x_m$ , where the literals $x_i$ correspond to variables in the combinational network in either complemented or uncomplemented form. Further the combinational network is modified as follows: We add an AND-gate with the output signal $f_t$ which becomes a new primary output of the combinational network. The inputs of the AND-gate are the variables of the combinational network corresponding to the literals in $t$. Inverters are added for those variables whose complements correspond to the literal. In other words, we implement the product term $t$ as an additional output of the combinational network. From the correctness and completeness of *make_all_implications()* and the definition of an implicant it follows that the assignment $y = 0$ implies $f_t = 0$ if and only if $t$ is a 1-implicant for $y$ (by contraposition $f_t = 1 \Rightarrow y = 1$.) Hence, $t$ is a 1-implicant if and only if for the above construction there exists an IST with identical leaves $f_t = 0$. To prove the theorem we show that without this construction an IST with leaves corresponding to the literals in $t$ exists if an IST with identical leaves $t = 0$ exists with the construction. For the construction, note that leaves of the IST must have siblings in the original AND/OR enumeration tree that correspond to the variables of $t$. This is guaranteed because $f_t = 0$ can only be implied from one of the inputs of the AND-gate and hence each leaf $f_t = 0$ of the IST with the construction must have a sibling in the original tree corresponding to a literal (variable) in the implicant. Hence, we can form an IST of the original tree which only contains literals of the implicant as leaves. Let $y$ be an arbitrary node in a combinational network and $T$ be the AND/OR enumeration tree for an initial set of value assignments $S = \{y = V\}$, $V \in \{0, 1\}$. For every *prime* implicant of $y$ there exists a minimal implication subtree (MIST) of $T$ such that the leaves of the MIST correspond to the literals of the prime implicant as is given in Theorem 4.2. Since $t$ is a prime implicant *all* literals of $t$ must be contained in the IST as leaves. The theorem is proved by observing that for any IST we can obtain a MIST as a subtree of the IST. The MIST is also an IST and hence its leaves must correspond to an implicant. Since the considered implicant is prime the MIST obtained from the IST must still correspond to the same implicant.

q.e.d.

**Theorem 4.4:** Let $y$ be an arbitrary node in a combinational network and $T$ be the D-AND/OR enumeration tree for a fault, $y$ stuck-at-1. Consider a product term $t = x_1 \cdot x_2 \cdot \ldots x_m$ where $x_i$ is a literal corresponding to a variable $f_i$ or its complement in the combinational network. Further, consider an IST of $T$ with a set of leaves $L$ such that in the combinational network the nodes $f_i$ *cannot be reached by the fault effect*.

If there is a one-to-one mapping between the literals $x_i$ of $t$ and the elements $(f_i = V_i)$ of $L$ such that $V_i = 0$ if $x_i$ represents the uncomplemented variable $f_i$, and $V_i = 1$ if $x_i$ represents the complemented variable $\overline{f_i}$, then $t$ is a permissible 1-implicant for $y$. Analogously, $t$ is a permissible 0-implicant for $y$ if the IST is a subtree of the enumeration tree for $y$ stuck-at-0.

**Proof:**

The proof is analogous to that of Theorem 4.2. Equations (1), (2) and (4) are still valid, because the terminal nodes ${}^{final}o_i$ are solely composed of variables which cannot be reached by the fault effect and thus can only take values $V \in \{0, 1\}$. The Boolean expression of Eq. (3) must be extended towards

$$(y = 0) \text{ and } (y \text{ observable at a primary output}) \Rightarrow {}^{0}a = 1.$$

Its contraposition states that if the Boolean expression evaluates to ${}^{0}a = 0$, $y$ is not observable or $y$ must be 1. We can therefore conclude that if $t$ evaluates to 1, $y$ is 1 or not observable. Hence, $t$ is a permissible 1-implicant. The proof for permissible 0-implicants is analogous.

q.e.d.

**Theorem 4.5:** Let $y$ be an arbitrary node in a combinational network and $T$ be the D-AND/OR enumeration tree for a fault, $y$ stuck-at-$V$, $V \in \{0, 1\}$. For *every permissible prime implicant* at node $y$ there exists a minimal implication subtree (MIST) of $T$ such that the leaves of the MIST correspond to the literals of the prime implicant as given in Theorem 4.4.

**Proof:**

The following is along the lines of the proof for Theorem 4.3. Routine *make_all_implications()* and *complete_unique_sensitization()* can identify all necessary assignments for single stuck-at fault detection at a node $y$ in a combinational network. These correspond to permissible single-literal implicants for $y$. This is accomplished by recursively checking whether all consistent justifications and sensitizations contain the same value assignments. Viewed in the AND/OR tree that corresponds to checking the properties of an IST according to Definition 4.5 such that the IST has identical leaves. Clearly, in the D-AND/OR enumeration tree permissible single literal implicants belong to implication subtrees with identical leaves.

The theorem is now proved for a permissible prime 1-implicant by the following construction: Suppose there is a product term $t = x_1 \cdot x_2 \cdot \ldots x_m$, where literals $x_i$ correspond to variables in the combinational network in either complemented or uncomplemented form. Further the combinational network is modified to add an AND-gate with the output signal $f_t$ which becomes a new primary output of the combinational network. The inputs of the AND-gate are the variables of the combinational network corresponding to literals in $t$. Inverters are added for those variables whose complements correspond to the literal. It follows from the correctness and completeness of routine *make_all_implications()* and the definition of an implicant that the assignment $y = 0$ and $y$ being observable implies $f_t = 0$ if and only if $t$ is a permissible 1-implicant for $y$ (by contraposition $f_t = 1 \Rightarrow y = 1$ or not observable). Hence, $t$ is a permissible 1-implicant if and only if for the above construction there exists an IST with identical leaves $f_t = 0$. It remains to be shown that without the above construction an IST with leaves corresponding to the literals in $t$ exists if an IST with identical leaves $t = 0$ exists with the construction. The rest of the proof is analogous to that of Theorem 4.3.

q.e.d.

**Theorem 5.5:** Let $y^i$ be a node of a combinational network $C^i$. The gates in this combinational network have no more than two inputs. Further, let $f^i$ be a divisor which is represented as combinational network and realizes a Boolean function of *no more than two* variables that may or may not be nodes in $C^i$ such that

1) The transformation of node $y^i$ into $y^{i+1}$ given by

$$y^{i+1} = f^i \cdot y^i + \overline{f}^i \cdot y^i$$

and followed by

2) Redundancy removal (with appropriate fault list)

generates a combinational network $C^{i+1}$. For an arbitrary pair of equivalent combinational networks $C$ and $C'$ there exists a sequence of combinational networks $C^1$, $C^2$,... $C^k$ such that $C^1 \equiv C$ and $C^k \equiv C'$.

**Proof:**

Switching algebra is isomorphic to two-valued Boolean algebra. A Boolean algebra can be defined by Huntington's axioms (see Section 1.1). First we show that all operations (transformations) defined by the axioms can be performed by the network manipulations given in the theorem. For each axiom it has to be shown that the corresponding transformation can be performed in both directions.

1) *Idempotent laws*: $a + a = a$; $a \cdot a = a$
for $a + a$:
$y = a + a, f = a$
Eq. 5.9→ $y = a \cdot (a + a) + \overline{a} \cdot (a + a)$ → redundancy elimination: $y = a \cdot (a + a)$ → redundancy elimination: $y = a$
$y = a,\ f = a + a$
Eq. 5.9 → $y = (a + a) \cdot a + \overline{(a+a)} \cdot a$ → redundancy elimination: $y = (a + a) \cdot a$ → redundancy elimination: $y = a + a$
for $a \cdot a$: analogous

2) *Commutative laws*: $a + b = b + a$; $a \cdot b = b \cdot a$
fulfilled by construction (definition) of primitive gates AND, OR.

3) *Associative laws*: $a + (b + c) = (a + b) + c$; $a \cdot (b \cdot c) = (a \cdot b) \cdot c$
for $a \cdot (b \cdot c) = (a \cdot b) \cdot c$:
$y = a \cdot (b \cdot c), f = c$
Eq. 5.9 → $y = c \cdot (a \cdot (b \cdot c)) + \overline{c} \cdot (a \cdot (b \cdot c))$ → redundancy elimination: $y = c \cdot (a \cdot (b \cdot c))$ → redundancy elimination: (because of commutative laws) $y = (a \cdot b) \cdot c$; in opposite direction analogous
for $a + (b + c) = (a + b) + c$: analogous.

4) *Absorption*: $a \cdot (a + b) = a$; $a + (a \cdot b) = a$

for: $a \cdot (a + b) = a$

$y = a \cdot (a + b), f = a$

Eq. 5.9 $\rightarrow y = a \cdot (a \cdot (a + b)) + \bar{a} \cdot (a \cdot (a + b))$ $\rightarrow$ redundancy elimination: $y = a \cdot (a \cdot (a + b)) \rightarrow$ redundancy elimination: $y = a$

$y = a, f = a + b$

Eq. 5.9 $\rightarrow y = (a + b) \cdot a + \overline{a+b} \cdot a \rightarrow$ redundancy elimination $\rightarrow y = (a + b) \cdot a$

for: $a + (a \cdot b) = a$ analogous

5) *Distributive laws*: $a \cdot (b + c) = (a \cdot b) + (a \cdot c)$; $a + (b \cdot c) = (a + b) \cdot (a + c)$

for: $a \cdot (b + c) = (a \cdot b) + (a \cdot c)$

$y = a \cdot (b + c), f = a \cdot b$

Eq. 5.9 $\rightarrow y = (a \cdot b) \cdot (a \cdot (b + c)) + \overline{ab} \cdot (a \cdot (b + c)) \rightarrow$ redundancy elimination: $y = a \cdot (b + c) + a \cdot b \rightarrow$ redundancy elimination: $y = (a \cdot b) + (a \cdot c)$

$y = (a \cdot b) + (a \cdot c), f = a$

Eq. 5.9 $\rightarrow y = a \cdot (a \cdot b + a \cdot c) + \bar{a} \cdot (a \cdot b + a \cdot c) \rightarrow$ redundancy elimination: $y = a \cdot (a \cdot b + a \cdot c)$ $\rightarrow$ redundancy elimination: $y = a \cdot (b + c)$

for: $a + (b \cdot c) = (a + b) \cdot (a + c)$ analogous

6) *Universal bounds*: $0 + a = a$; $0 \cdot a = 0$; $1 + a = 1$; $1 \cdot a = a$

for $0 + a = a$:

$y = 0 + a, f = a$

Eq. 5.9 $\rightarrow y = a \cdot (0 + a) + \bar{a} \cdot (0 + a) \rightarrow$ redundancy elimination: $y = a \cdot (0 + a) \rightarrow y = a$

$y = a, f = 1$

Eq. 5.9 $\rightarrow y = 1 \cdot a + 0 \cdot a \rightarrow$ redundancy elimination for stuck-at-1 fault at $a$ in second summand: $y = 1 \cdot a + 0 \rightarrow$ redundancy elimination for stuck-at-1 at signal with constant one: $y = a + 0$

7) *Unary operation*: $a \cdot \bar{a} = 0$; $a + \bar{a} = 1$

for: $a + \bar{a} = 1$

$y = a + \bar{a}, f = 1$

Eq. 5.9$\rightarrow y = 1 \cdot (a + \bar{a}) + 0 \cdot (a + \bar{a}) \rightarrow$ redundancy elimination: $y = a + \bar{a} \rightarrow$ redundancy elimination: 1

$y = 1, f = a + \bar{a}$

Eq. 5.9 $\rightarrow (a + \bar{a}) \cdot 1 + \bar{a} \cdot a \cdot (1) \rightarrow$ redundancy elimination: $y = a + \bar{a}$

In order to complete the proof it must be shown that the expansion given in the theorem also allows arbitrary sharing of logic. This follows easily from the following construction. Let $C$ be the original network and $C'$ be the target network. Further, let $C_{tree}$ denote a network that has tree structure and results from $C$ if all sharing of logic is removed by duplication. Similarly, let $C_{tree}'$ denote the tree version of the target network. Consider the following construction: Remove all sharing of logic between the different output cones of the original network $C$ so that we obtain $C_{tree}$. It is easy to derive $C_{tree}$ by the given expansion. Let $y$ be some internal fanout point and assume it is the output of an AND gate with input signals $a$ and $b$. By choosing a divisor $f = a \cdot b$ and by performing the given expansion with appropriate fault list, the fanout point is moved to the inputs of the AND gate. For other gate types the procedure is analogous. This process is repeated until no more internal fanout points exist and $C_{tree}$ has been obtained. After all sharing of logic has been removed each output cone is isomorphic to a Boolean expression that can be manipulated arbitrarily as shown using the above axioms. Therefore, it is also possible to obtain the network $C_{tree}'$ by the given expansion. The target network $C'$ results if the duplicated logic is removed. This can be accomplished if equivalent nodes are substituted. If node $y$ is to be substituted by node $y'$ this can be accomplished by selecting $f = y'$ and performing the given expansion. This process can be repeated for well-selected nodes in $C_{tree}'$ until network $C'$ is reached.

q.e.d.

**Lemma 5.1:** Consider a function: $y' = y|_1 + \bar{f}$. Then, $y' = y$ if and only if the implication $y = 0 \Rightarrow f = 1$ is true.

**Proof:**

$\Rightarrow$:

$(y = 0 \Rightarrow f = 1) \Leftrightarrow (f = 0 \Rightarrow y = 1) \Rightarrow$ in Eq. 5.7 $y|_0$ can be set to 1 and we obtain: $f \cdot y|_1 + \bar{f} \cdot y|_0 = f \cdot y|_1 + \bar{f} = y|_1 + \bar{f}$

$\Leftarrow$:

$y|_1 + \bar{f} = f \cdot y|_1 + \bar{f} \cdot y|_0$ (Eq. *)

$\Rightarrow \quad f \cdot (y|_1 + \bar{f}) = \bar{f} \cdot (f \cdot y|_1 + \bar{f} \cdot y|_0) \Leftrightarrow \bar{f} = \bar{f} \cdot y|_0 \Rightarrow f + \bar{f} = f + \bar{f} \cdot y|_0$

$\Leftrightarrow \quad 1 = f + y|_0$

$\Rightarrow$ for $f = 0$ it follows that $y|_0 = 1$, for $f = 0$ the function $y|_0$ assumes the same values as $y$ (by definition), hence the implication $f = 0 \Rightarrow y = 1$ must be true and by contraposition the implication $y = 0 => f = 1$ must be true if Eq. * can be fulfilled

q.e.d

**Theorem 5.6:** Let $y^i$ be a node of a combinational network $C^i$. The gates in the combinational network can have no more than two inputs. Further, let $f^i$ be an implicant according to Definition 4.3 such that

1) the transformation of node $y^i$ into $y^{i+1}$ given by

$$y^{i+1} = y^i + \overline{f}^i \quad \text{for } y = 0 \Rightarrow f = 1$$
$$y^{i+1} = f^i + y^i \quad \text{for } y = 0 \Rightarrow f = 0$$
$$y^{i+1} = f^i \cdot y^i \quad \text{for } y = 1 \Rightarrow f = 1$$
$$y^{i+1} = \overline{f}^i \cdot y^i \quad \text{for } y = 1 \Rightarrow f = 0$$

followed by

2) redundancy removal (with appropriate fault list)

generates a combinational network $C^{i+1}$. For an arbitrary pair of equivalent combinational networks $C$ and $C'$ there exists a sequence of combinational networks $C^1$, $C^2$,... $C^k$ such that $C^1 \equiv C$ and $C^k \equiv C'$.

**Proof:**

Follows from the proof of Theorem 5.5, by noting that all functions $y$ and divisors $f$ used in that proof satisfy one of the implications specified in the theorem.

q.e.d.

**Theorem 5.7:** Let $f$ and $y$ be arbitrary nodes in a combinational network $C$ where $f$ is not in the transitive fanout of $y$ and both stuck-at faults at node $y$ are testable.

The function $y'$: $B_2^n \rightarrow B_2$ , $B_2 = \{0, 1\}$ with $y' = y|_1 + \bar{f}$ is a *permissible function* at node $y$ if and only if the D-implication $y = 0 \xrightarrow{D} f = 1$ is true.

**Proof:**

$\Rightarrow$:

If $y = 0 \Rightarrow f = 1$ is true, then $y = 0 \xrightarrow{D} f = 1$ is true. We partition the set of possible combinations of input assignments (rows in the truth table) into two disjoint subsets, where each fulfills one of the following conditions:

Case 1: ($y = 0$ and $f = 1$) or $y = 1$:
For these inputs $y = 0 \Rightarrow f = 1$ is true and Lemma 5.1 applies, for these inputs $y$ and $y'$ always assume the same value.

Case 2: $y = 0$ and $f = 0$:
For these inputs $y'$ can have a different value than $y$. With $y = 0$ the function $y'$ can only assume the faulty value, 1. However, this cannot lead to a wrong value at the primary outputs of $C$ because the fault stuck-at-1 at node $y$ cannot be tested as $f = 0$ and $y = 0 \xrightarrow{D} f=1$.

$\Leftarrow$:

The transformation considered in the theorem is permissible if one of the following cases is fulfilled:

Case 1: $y$ and $y'$ are equivalent
applies, if $y = 0 \Rightarrow f = 1$ is true then $y = 0 \xrightarrow{D} f = 1$ is true.

Case 2: $y = 1$ and $y' = 0$ and $y'$ stuck-at-0 is untestable under this condition
$y = 1 \Rightarrow y|_1 + \overline{f} = y' = 1$, i.e., this case cannot occur.

Case 3: $y = 0$ and $y' = 1$ and $y'$ stuck-at-1 is untestable under this condition. With $y' = y|_1 + \overline{f}$, $y = 0$ and $y' = 1$ can only occur if $f = 0$ is true. The term $y|_1$ in the given transformation means this node can be implemented as an arbitrary function assuming the same values as $y$ for $f = 1$. As a special case assume that $y|_1 = y$. If $y = 0$ then $f = 0$ is sufficient to produce a "faulty" signal, 1, at node $y'$. Now consider the set of all test vectors for $y$ stuck-at-1 in the original circuit that produce $f = 0$. Every such test will result in a faulty response of the transformed circuit. Therefore, the transformation is only allowed if such a test does not exist. However, if a test for $y$ stuck-at-1 exists, in general, it is required that there is none which produces $f = 0$. This means that, $f = 1$ is necessary for fault detection and $y = 0 \xrightarrow{D}$ $f = 1$ must be true. If this condition is necessary for the special case $y|_1 = y$, it is also necessary for the general statement since $y$ is one of the possible choices to implement $y|_1$.

q.e.d

# INDEX


## A

aborted fault 34
aborting criterion 34
absorption 3
algebraic techniques 4, 108
alternative graph 15
AND/OR enumeration 42, 55, 58
AND/OR reasoning 52
AND/OR reasoning graph 15, 49, 75
AND/OR reasoning tree 82, 86
AND/OR search 77
AND/OR tree 85
*and_or_based_variable_selection()* 149
*and_or_enumerate()* 84
apply 14, 181
associativity 3
ATPG 17
automatic test pattern generation 17

## B

backtrace 33, 34
backtracking 22
BDD 11
binary decision diagram 11, 78
  reduced, ordered 11
BOLD 118
Boolean algebra 1, 3
Boolean difference 29
Boolean expression 6
Boolean function 5
Boolean network 10
Boolean representation 5
Boolean space 5
Boolean techniques 108
bound set 104
branch-and-bound 22
branching program 11

## C

canonical form 9, 11
Cartesian product 1
circuit netlist 10
circuit partitioning 166
cofactor 132
co-kernel 113
column multiplicity 106
combinational circuit 10
combinational network 10
commutativity 3
complement 3
complete set of justifications 57
*complete_unique_sensitization()* 65
compose 14, 189
conflict 33
conjunction 3
conjunctive normal form 8
connection 103
connection count 102
*consistent_satisfy()* 183
contraposition 45
controllability 29, 33
controllability don't cares 119
controlling set 121
controlling value 36
COP 33
cover 9
cube 109
cube-free 113
CUT 18
cycle 4

## D

DAG 4
D-calculus 23
decision tree 21
decomposition 103
  chart 104

disjoint 104
simple 104
strict 107
*demo_recursive_learning()* 50
design for testability 18
deterministic ATPG 18
D-frontier 30
D-implication 139, 168
disjunction 3
disjunctive normal form 8
division 108, 133
algebraic 109
Boolean 109, 111
divisor 133
dominator 37
don't care set 5
internal 119, 133

E

EC 164
engineering changes 164
ESPRESSO 9, 113, 118, 130
event list 58
explicit enumeration 23
extraction 117

F

factored form 6
factoring 116
false negative 168
FAN 37
fanin 4
transitive 5
fanout 4
transitive 5
fault coverage 19
fault dropping 21
fault injection 29
fault list 18
fault model 18
fault propagation 35
fault signal 29
*find_implicant()* 150
fixed logic value 26
forcing set 121
FPGA 104
free set 104
FSIM 154
function in combinational network 10

G

gate 6
gate netlist 10
global flow 120, 144
graph 4
acyclic 4
directed 4

H

HANNIBAL 154
hashing 97
Huntington's postulates 3

I

idempotency 3
if-then-else-operator 14
implicant 9, 82
in Boolean network 90
multi-literal 91
permissible 94
prime 9, 91
single literal 91
implication 26, 31, 55
backward 26
complete 56
direct 26
forward 26
global 32
indirect 27
precise 56
simple 26
implication subtree 92
implicit enumeration 23
*include_don't_cares()* 186
incremental synthesis 164
indegree 4
irredundant 129
IST 92

## J

J-frontier 30
justification 50, 57

## K

kernel 113
  level 115

## L

lattice 2
learning 42, 55
  criterion 44
  dynamic 42
  recursive 49, 55, 139
  rule 44
  static 42
LEVEL 33
library 10
literal 6
literal count 102
literal subtree 146
logic alphabet 7
logic consistency 32
logic optimization 101
logic value 7
logic verification 163, 197
LSS 121
LST 146
LUT 104

## M

*make_all_implications()* 58
maxterm 8
minimal implication subtree 93
minterm 8
MIS 113
MIST 93
miter 169

## N

necessary assignment 29, 31, 55
negation 3
node in combinational network 10
non-controlling value 36
non-solution area 39

## O

OBDD 11
objectives in ATPG 34
observability 33
observability don't cares 94, 120, 138
off-set 5
  maxterm 8
  minterm 8
on-set 5
  maxterm 8
  minterm 8
optional assignment 31, 33
OR graph 78
OR search 77
ordered pair 1
orthonormal expansion 9, 130, 132
outdegree 4

## P

partially ordered set 2
path, directed 4
permissible function 94, 139, 168, 224
phase assignment 103
PODEM 38
predecessor 4, 5
primary inputs 10
primary outputs 10
prime 129
product term 8
product, boolean 2

## Q

Quine-McCluskey method 9, 129
quotient 133

## R

RAMBO 154
reachable by fault 36

reasoning 81
recursion depth 67
reduce 13
redundancy elimination 125
redundant fault 19, 34
relation 1
remainder 133
resolution 77
restrict 14, 132
resubstitution 118
ROBDD 11

S

satisfiability 82
satisfiability problem 11
satisfying set 183
  consistent 183
SCOAP 33
search 77
Shannon tree 78
Shannon's expansion 9, 132
signal in combinational network 10
similarity 166
simplification 119
SIS 157
specified logic value 26
STAFAN 33
stuck-at fault 18
subgraph 4
substitution 118
successor 4, 5
sum of product 9
  irredundant 9
  prime 9
sum term 8
sum, boolean 2

T

target fault 18
tautology 87
test vector 19
testability measure 33
TOPS 37
transitive closure 47, 60
tree
  directed, rooted 4
truth table 7

U

unate SOP 9, 87
unique sensitization 31, 35, 37, 55
universal bounds 3
unjustified gate 55
unjustified lines 30
unspecified logic value 26

V

variable in combinational network 10
variable ordering 15

X

X-path 32